To:
  ILARIA
  COSIMO and
  FRANCESCA

EVOLUTION
of the
PRIMATES
An Introduction to the
Biology of Man

A. B. CHIARELLI

Institute of Anthropology, Primatology Centre
University of Turin, Italy

1973

ACADEMIC PRESS
London and New York

ACADEMIC PRESS INC. (LONDON) LTD
24–28 Oval Road
London NW1

U.S. Edition published by
ACADEMIC PRESS INC.
111 Fifth Avenue,
New York, New York 10003

Library of Congress Catalog Card Number: 73–141739
ISBN: 0–12–172540–5

PRINTED IN GREAT BRITAIN BY:
W & J Mackay Limited, Chatham

Preface

This book is based on a series of lectures delivered in a course on Anthropology and Primatology at the University of Turin. My notes were subsequently translated into English, and I delivered a new set of lectures on "Problems of Human Evolution" in the Department of Anthropology, University of Toronto, where I had the honour of being a guest lecturer in 1971 and 1972.

Many thanks are due to students who attended the courses both in Turin and Toronto, and whose suggestions were responsible for much revision and up-dating of the text.

The book contains standard facts about primate evolution and the origin of Man in as organic a form as possible. I was aware, however, that there is a certain lack of depth inherent in this approach. Yet my main concern was to furnish a basis for the understanding of evolutionary phenomena, and to add to this basis the newest data and findings relevant to the various aspects of primate evolution. Many data relate to research on living species that is now being carried out.

Information on fossil remains is preceded by a critical survey of the reliability of such information as well as a breakdown of the climatic factors that we can now deduce existed and influenced the earth during the second half of the Tertiary and in the Quaternary.

There are many aspects of the differences and evolutionary changes between anthropoid apes and humans that have been specifically outlined, i.e. evolution of erect posture, opposability of the thumb, cranial capacity and evolution of the brain, etc.

I have intentionally disregarded the social organization, behaviour and ethology of primates, although I believe these things to be of fundamental importance, because I wanted to concentrate on physical evolution.

But in order to offer a balanced and comprehensive synthesis of a variety of subjects, sometimes I have given only cursory attention to details, offering instead numerous general reference lists which the student can use.

Some chapters are derived in conception from the works of eminent scholars such as Campbell, Napier and Delattre who have treated the particular subjects in infinitely greater depth, and to whom I am indebted for inspiration.

Many grateful thanks are due to all those who helped in the achievement of such a laborious undertaking; in particular to my translators Mrs. J.

Raikes and Miss Rory Daniel; to Miss Clelia Kostas and Miss Maria Monchietto for their excellent drawings, to my secretary, Miss Chiara Bullo who typed the manuscript, to my students in Turin and Toronto, and to the production department of Academic Press.

Last, but not least, I wish to acknowledge my wife's endurance for all the time I had to steal from her.

Turin, November 1972   A. B. CHIARELLI

Contents

5 The Living Primates

6 Evolution of Species (Speciation)

7 Comparison of the Chemical Compositions of Single Molecules in Primates

8 Comparison of Single Hereditary Characters of Man with Those of Other Primates

9 A Comparison between Serum Proteins and Immunological Distances

10 The Genetic Bases of Morphological Characters

11 The Karyotype of the Primates and of Man

12 Palaeontological Data as Proof of the Differentiation of Species

13 Time in the Evolution of the Primates and of Man: Methods of Measuring it and its Subdivisions

14 Climate during the Period of the Evolution of the Primates and the Appearance of Man

15 Fossil Primates

16 Hominoid and Hominid Fossils

17 Posture of the Primates and the Acquisition of an Upright Posture by Man

18 Differentiation of the Upper and Inferior Limb in the Primates and the Opposability of the Thumb in Man

19 The Evolution of the Head in Relation to Erect Posture

20 Cranial Capacity and Its Growth from Primates to Man: the Differentiation and Evolution of the Brain

I

Primate Evolution and the Natural History of Man

The Primates are a particularly fortunate and successful order of outstanding interest, unique among mammals. Considered as a zoological group, all the living forms give examples of evolutionary organization which broadly duplicate the evolutionary stages leading to man. In fact, prosimians, monkeys and anthropoid apes may represent the survival of stages reached in the past by human ancestors.

The horizontal classification duplicates, in its major features, the vertical: that of the fossil forms. For this reason the group is of particular interest to the zoologist.

However, the study of the primates is also of great human interest, for this is, in fact, the Order to which we ourselves belong. The resemblance to our species, not just in external appearance but also in behaviour, of the monkeys, in general, and the anthropoid apes, in particular, cannot be missed by anyone familiar with the higher animals nor even by the most casual visitor to a zoo.

Awareness of this similarity is very ancient, and so too is the documentation testifying to it. The first written account is probably that related by Hanno in his Journal, the "Periplus Hannonis" (470 B.C.); the following is a passage from this document: "We arrived at an inlet which is called the Southern Horn; further off there was another island populated by savages. The majority of them were of the female sex, their entire bodies covered with hair, and the interpreter called them 'gorillas'. It was not possible to approach the males, whom we at once made ready to pursue. They escaped us, climbing nimbly up the cliffs and defending themselves with stones. After a great effort we succeeded in capturing three females, whom, however, it was impossible to carry on board because they bit and scratched; we were forced to kill them and carry their skins to Carthage."

This awareness is also emphatically clear in the writings of Aristotle (384–322 B.C.). *A propos* of the monkeys in his "Historia Animalium" he propounds his general principle of the gradualness of nature (*Natura non facit saltus*) in these words: "there are some animals, such as the monkeys and the baboons, which, because of their ambiguous nature, have both hands and feet, but their feet can be used as hands."

However, among the ancients it was Galen (A.D. 131–200) who examined these similarities in detail at the anatomical level. "The monkeys are, among the animals, those which most resemble man in their viscera, muscles, arteries, nerves and bones . . ." Such were the convictions propounded by the philosophy and science of those days. Man was looked upon as an animal resembling the monkeys, distinguished from them by the larger size of his brain, by his intelligence and by his upright carriage.

After the long interval of the Middle Ages, with the revival of interest in natural science this concern with our likeness to the monkeys and with the animal nature of man was also renewed and deepened. The work of Edward Tyson (1650–1708), the first real treatise on the comparative anatomy of man and the chimpanzee, dates from 1699.

At the end of the seventeenth century this naturalistic vision of man, that is the unqualified inclusion of our species in the animal kingdom, found its systematist in Carolus Linnaeus (1707–1778), the famous founder of the system of binominal nomenclature. In his "Systema Naturae" he divides animals into three groups: primates, secundates and tertiates. Among the primates he includes man and the monkeys; among the secundates all the rest of the mammals; and among the tertiates all other animals.

Moreover, in his classification of the primates, Linnaeus ascribes the name *Homo* not only to our species but to the chimpanzee and orangutan as well, distinguishing them with the attributes *sapiens*, *sylvestris* and *troglodytes* respectively. He justifies this action in the following phrase: "Although I have given much attention to it, I have not been able to discover any difference between man and the chimpanzee." Later he says, "As a naturalist, and following naturalistic methods, I have not been able to discover up to the present a single character which distinguishes Man from the anthropomorphs, since they comprise specimens that are even less hairy than Man, that walk erect on two feet and that resemble the human species in their hair and their hands to such a degree that an inexpert traveller may consider them varieties of men. But there is in Man something involving a consciousness of himself, which is reason."

Such was the common knowledge of this likeness that monkeys, in the illustrations in books of the period, were always represented with a very human countenance. It was in this way that students of Linnaeus, used to depict the monkeys then known (Fig. 1.1).

However it was not until the publication of the "Origin of Species" (1859), or more exactly of "The Descent of Man" (1871) by Darwin, that the question of the relationship between men and monkeys was clearly stated and became inflammatory and dramatic. Everyone is familiar with the many disputes which originated from this problem.

We do not want to become involved in an over-lengthy discussion of

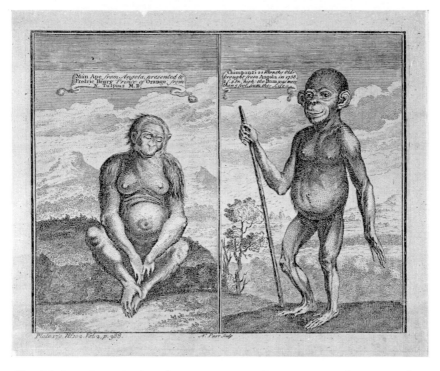

FIG. 1.1 Apes drawn to give a human appearance from a seventeenth-century print.

how much of this discord was the result of a genuine misinterpretation of the terms of the question and how much, on the other hand, it was the fruit of the anthropocentric, idealistic and teleological philosophy particularly favoured at that time. The fact is that from the publication of Darwin's book there arose among scholars of different disciplines violent arguments and controversies which have died away only very recently with the complete acceptance of the real and close biological relationship between our species and the other primates.

Nevertheless the philosophy of man as the centre of the universe, of man as a superior form of life, continues to be bruited in its most damaging terms. It may be that this is the source of many of the evils that constantly afflict all of mankind: the continual presumption of being superior, and of not only knowing what is true but of being the sole guardians of the truth. Many people think that it is necessary and important to reverse this process of mythicizing man in order to replace him in his true position within the scheme of Nature. This could lead to a new form of Humanism in which scientific knowledge (rather than rights and charity) would underlie the laws governing the relationships between men, and between men and things.

If it is acknowledged that this concept is correct, is it likely that the documentation and the promulgation of the knowledge of the true nature of our species would lead to the reversal of this anthropocentric process? It is doubtful. Nevertheless it is certainly potentially useful to present yet again to Humanity as it exists verification of its real nature as an animal species. Human beings must be reminded that they are animals whose lives unfold in a mere instant of time; that their survival is secured by that of other individuals as part of a community; that they are enclosed in a physical and biological environment which, subjected to constant unnatural transformations, must be consciously respected, and of which all aspects, including a proper numerical equilibrium, must be maintained.

A second reason why the study of Primate evolution can be considered useful is concerned with its role in the systematization of the knowledge that Mankind has so far acquired. The students of Man, the anthropologists, are mainly concerned with describing the various characteristics and peculiarities of the human species, just as the zoologists do for a particular animal species, or the botanists for plants. Obviously methods will differ according to the preparation and interests of each scientist, but the results aim at the construction of a naturalistic synthesis of all data obtained. Such a synthesis ultimately must form a bridge between what have unfortunately come to be called the "Experimental Sciences" and the "Humanistic Sciences".

If moreover we wish to place this science (Anthropology) in a wider framework, we must first attempt to divide the story of the world into three broad periods: that of chemiogenesis, biogenesis and cognogenesis.

The chemiogenetic period is characterized by the production of complex organic compounds which originated by means of mechanisms such as primitive cosmic unions, photochemical processes, heat reactions or spontaneous reactions between preformed compounds. However, these were not reproductive processes as the compounds were not formed by the replication of identical structures.

The biogenetic period is characterized by the reproduction of polymers, built up according to a specific order. The biogenetic polymer *par excellence* of our globe is deoxyribonucleic acid (DNA). This molecule determines the sequence of its own reproduction and of those materials such as proteins which constitute cells and organisms. Errors in reduplication of this polymer have occurred by chance in nature, and natural selection has chosen among the best adapted, thus creating the enormous variety of life forms existing on the earth today.

The cognogenetic period (which could be called History) is characterized by the evolution of the mechanisms of perception and calculation as well as by the use of abstract symbols and interpersonal communication, all of which were necessary for the accumulation of tradition and of culture.

Anthropology is concerned with the study of the biogenesis of the primate groups, the physical origin of Man, and the origin and the mechanisms of evolution in the primitive cognogenetic period.

However, even if these zoological and humanistic interests do not command serious consideration, there are at least two other points of view from which the study of the primates merits attention. Firstly, the established biological relationship between our species and those of the other primates suggests that the study of the comparative biology of the non-human primates will be the precursor of new and important discoveries in human biology. Secondly, the indisputable observation that, on the basis of many biological criteria, man is by far one of the most advanced products of organic evolution, compels our interest in how this exceptional standard was achieved.

With these points in mind we shall, in the following pages, attempt to document the facts of the evolution of man by means of comparisons with the types of animals most directly related to him. A proper understanding of human evolution must, in fact, be based upon a detailed knowledge of the evolutionary trends within the entire order of primates.

The fundamental unit for any systematic study of a group of living organisms and their evolution is the individual. Therefore we shall begin by trying to determine the characteristics of a typical primate individual, and for convenience we shall examine a representative of our own species. Every individual can be described by his external appearance, by his measurements or by a variety of other features. We wish to portray our chosen individual from the genetic point of view: since the study of evolution is based essentially on the methods of genetics.

GENERAL REFERENCES

Beals, R. L. and Hoijer, H. (1965). "An Introduction to Anthropology", Mac-Millan Co.; Collier-MacMillan, London and New York.

Buettner-Janusch, J. (1966). "Origins of Man, Physical Anthropology", J. Wiley & Sons, New York, London, Sydney.

Campbell, B. G. (1966). "Human Evolution: An Introduction to Man's Adaptations", Aldine Publishing Company, Chicago.

Comas, J. (1962). "Manual de Antropología Física", Universidad Nacional Autónoma de México, Instituto de Investigaciones históricas, Sección de Antropología, Mexico.

Lasker, G. W. (ed.) (1960). "The Processes of Ongoing Human Evolution", Wayne State University Press, Detroit.

Lasker, G. W. (1961). "The Evolution of Man. A Brief Introduction to Physical Anthropology", Holt, Rinehart & Winston, New York and London.

Laughlin, W. S. and Osborne, R. H. (eds) (1968). "Human Variation and Origins. An Introduction to Human Biology and Evolution", W. H. Freeman & Co., San Francisco and London.

Montagu, A. (1960). "Introduction to Physical Anthropology", Charles C. Thomas, Springfield, Illinois, U.S.A.

Napier, J. (1971). "The Roots of Mankind", G. Allen and Unwin, London.

2

The Genetic Aspect of Evolution

1. GENETIC CHARACTER OF THE INDIVIDUAL

Other than in the exceptional case of monozygous twins, two almost identical individuals do not exist, and probably never have existed in our own or any other primate species. Differences are not limited to visible characteristics such as eye or hair colour or nose shape, but include those traits perceptible by senses other than sight and those which our senses cannot discern at all. Every individual differs from others by a particular tone and timbre of voice and is, moreover, characterized by a definite odour which we rarely notice but by which our dogs can recognize us. Each of us is characterized by a particular blood group which can be demonstrated only by special laboratory techniques. These differences have aroused interest from ancient times: they have been accepted fatalistically or bemoaned as unjust; they have constituted motives for reverence or causes for persecution; they have been attributed to heredity or to environment, to ancestry or to education. Often the facts have been misrepresented and, more or less deliberately, erroneous conclusions have been drawn from them.

However, notwithstanding this enormous variability, each of us considers himself a human being and distinguishes himself and his fellow humans from other living things, from which it must be concluded that all the individuals belonging to our species possess definite characteristics in common. Our distinctions from other animals are most obvious when we consider characteristics which appeared farthest back in the course of evolution. Characteristics appearing more recently are generally of minor importance and their hereditary nature is generally demonstrable.

Human beings are chordates inasmuch as in one of the first stages of development we possess a supporting structure arranged longitudinally in the dorsal aspect of the alimentary canal. We assume that this is also a genetically-determined character, but are unable to determine how many or which genes are involved in the formation of the notochord. Then we must also classify ourselves among the vertebrates, since we possess an ossified skeleton. No mutations are known which eliminate the skeleton, but a hundred more or less small hereditary variations in its organization are known, which in varying ways and to different degrees interfere with the normal functions of life.

Our mammalian characteristics, i.e. having a hairy covering, being provided with mammary glands and having a constant body temperature, are sometimes less stable. Different mutations can lead to a more or less complete absence of hair and to variations in the number, in the location and in the functional ability of the mammary glands. In these cases unfavourable mutations can be compensated for by our ability as human beings to control our environment. We can, in fact, protect ourselves from the cold with clothing; we can encourage the growth of mammary glands by the administration of hormones, and we can raise infants on artificial feedings.

Proceeding along these lines, we shall ultimately be able to recognize a certain number of characters which distinguish the species *Homo sapiens* from the anthropoid apes and the latter from the other species of primates. Moreover, within the species *Homo sapiens*, by means of geographical and social isolation and with the concomitant effects of crossing, of genetic drift and of selection, many populations have been differentiated from others by one or more characteristics. These are sometimes conspicuous such as eye or skin colour, hair form or stature.

Finally, within the limits of one population, the environmental and genetic differences which determine the physical aspect of an individual are so powerful and numerous that each individual must be considered unique. The uniqueness of every individual of a population is evident when the smoothly functioning complexity of his genotype is considered, and even more clearly apparent when the complicated organization of each of his molecules is studied in detail.

2. IMMUTABILITY OF THE SUCCESSION OF THE AMINO ACIDS IN A PROTEIN

Proteins are one of the essential constituents of our bodies. They are molecules having very high molecular weights, and are largely made up of a succession of amino acids (Table 2.1). Their number and order is fixed for each of the myriad of different proteins which are concerned in making up each of our bodies (Fig. 2.1).

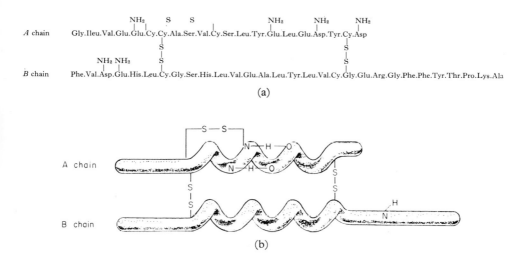

(a)

(b)

FIG. 2.1 The structure of insulin (a) (from A. P. Ryle *et al.*, *Biochem. J.* 60, 541 (1955), and (b) a schematic drawing of how insulin might be coiled within the portion of the molecule which is stabilized by disulphide bridges.

The architecture of each protein molecule is determined *a priori* by a genetic code organized in the deoxyribonucleic acid (DNA) found in the chromosomes. This code, aided by messenger RNA (ribonucleic acid) and by insoluble RNA located in the ribosomes, determines the construction of the protein chain.

The information necessary for protein synthesis is codified in the sequence of fundamental subunits (nucleotides) which have the names of the 4 bases which compose them: adenine, guanine, cytosine, thymine. These are the letters necessary for the formation of the "syllables" and hence, the "words" necessary to read the genetic message. Since it is not possible to have an unequivocal correspondence between the 4 bases and the 20 amino acids, only a sequence of 3 bases (triplets) can indicate which amino acid must be incorporated into a protein chain.

Table 2.1. Chart of twenty naturally occurring amino acids which are the constituents of proteins.

Amino acid	Structure (R group)
	$O=C(HO)-\overset{\overset{H}{HN}}{\underset{H}{C}}-R$ (general structure)
Glycine (Gly)	$-\overset{|}{\underset{H}{C}}-H$
Alanine (Ala)	$-\overset{|}{\underset{H}{C}}-CH_3$
Valine (Val)	$-\overset{|}{\underset{H}{C}}-CH-CH_3$; CH_3
Isoleucine (Ileu)	$-\overset{|}{\underset{H}{C}}-CH-CH_2-CH_3$; CH_3
Leucine (Leu)	$-\overset{|}{\underset{H}{C}}-CH_2-CH-CH_3$; CH_3
Phenylalanine (Phe)	$-\overset{|}{\underset{H}{C}}-CH_2-$ ⬡
Tyrosine (Tyr)	$-\overset{|}{\underset{H}{C}}-CH_2-$ ⬡ $-OH$
Cysteine (Cys)	$-\overset{|}{\underset{H}{C}}-CH_2SH$
Methionine (Met)	$-\overset{|}{\underset{H}{C}}-CH_2-CH_2-S-CH_3$
Serine (Ser)	$-\overset{|}{\underset{H}{C}}-CH_2OH$
Threonine (Thr)	$-\overset{|}{\underset{H}{C}}-CH-CH_3$; OH
Aspartic Acid (Asp)	$-\overset{|}{\underset{H}{C}}-CH_2-C\overset{O}{\underset{OH}{}}$
Asparagine (Asp-NH$_2$)	$-\overset{|}{\underset{H}{C}}-CH_2-C\overset{O}{\underset{NH_2}{}}$
Glutamic Acid (Glu)	$-\overset{|}{\underset{H}{C}}-CH_2-CH_2-C\overset{O}{\underset{OH}{}}$
Glutamine (Glu-NH$_2$)	$-\overset{|}{\underset{H}{C}}-CH_2-CH_2-C\overset{O}{\underset{NH_2}{}}$
Lysine (Lys)	$-\overset{|}{\underset{H}{C}}-CH_2-CH_2-CH_2-CH_2-NH_2$
Arginine (Arg)	$-\overset{|}{\underset{H}{C}}-CH_2-CH_2-CH_2-NH-C\overset{NH}{\underset{NH_2}{}}$
Histidine (His)	$-\overset{|}{\underset{H}{C}}-CH_2-C=CH$; HN , N , CH
Proline (Pro)	$-\overset{|}{\underset{H}{C}}-\overset{\overset{H}{N}-CH_2}{\underset{CH_2}{}}CH_2$
Tryptophan (Try)	$-\overset{|}{\underset{H}{C}}-CH_2-C$ (indole ring)

Asparagine = Asn; Glutamine = Gln.

TABLE 2.2. The genetic code. In this table the first letter of each triplet is read in the vertical column of the left, the second in the upper line, the third in the vertical column on the right.

1°\2°	U	C	A	G	3°
U	PHE	SER	TYR	CYS	U
	PHE	SER	TYR	CYS	C
	LEU	SER	nonsense	nonsense	A
	LEU	SER	nonsense	TRY	G
C	LEU	PRO	HIS	ARG	U
	LEU	PRO	HIS	ARG	C
	LEU	PRO	GLU-NH$_2$	ARG	A
	LEU	PRO	GLU-NH$_2$	ARG	G
A	ILEU	THR	ASP-NH$_2$	SER	U
	ILEU	THR	ASP-NH$_2$	SER	C
	ILEU	THR	LYS	ARG	A
	MET	THR	LYS	ARG	G
G	VAL	ALA	ASP	GLY	U
	VAL	ALA	ASP	GLY	C
	VAL	ALA	GLU	GLY	A
	VAL	ALA	GLU	GLY	G

The number of possible triplets, obtained by the permutations of the 4 bases, is 64 (Table 2.2), while there are only 20 amino acids. Since the first studies of Niremberg and Mattaei an explanation of this apparent "waste" of letters has been found so that we now know the meaning of all 64 triplets. One of the first steps in the deciphering of the code was the use of very simple synthetic polyribonucleotides. In this way it was seen how poly-U is capable of codifying phenylalanine, poly-C proline, etc.; to the point where by means of the study of ever more complex polyribonucleotides, it was possible to find the key to the code.

The contradiction between the number of syllables and the number of amino acids has been explained by showing that some amino acids are codified by more than one triplet of nucleotides (UUU and UUC both code phenylalanine), that some triplets do not have any meaning (UAA, UAG), and that others signal the beginning or the end of a chain of amino acids.

The syllables read on the DNA are transcribed on molecules of RNA synthesized in proximity to the chromosomal DNA. The letters of the alphabet of RNA are also 4 in number; three are the same as those of DNA (A, G, C,) while the fourth is different (uracil being substituted for thymine).

The message thus transcribed from one alphabet to another is transferred from the nucleus to the cytoplasm of the cell by "messenger" RNA (mRNA). This is attached to a ribosome (the real agent of protein synthesis) and is read in a fixed direction so that a new amino acid is joined to each triplet, thus translating the language expressed in nucleotides into a language of amino acids. In this operation another type of RNA called transfer RNA (tRNA) comes into play. There is a different type of tRNA for each amino acid so that it will tally with the different sequence of nucleotides in the transcript.

Each tRNA selects the correct amino acid in the cytoplasm and inserts it in the right place in the protein chain being formed. When the ribosome has read all the message written on the RNA it separates itself from it and from the completed protein and is ready for another message. The entire process lasts only a few seconds (Fig. 2.2).

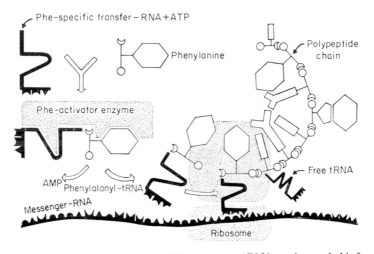

FIG. 2.2 Mechanism of protein synthesis. The messenger RNA receives coded information from the DNA in order to carry out protein synthesis.

3. WHAT IS A GENE AND HOW MANY GENES ARE THERE IN THE NUCLEUS OF A HUMAN CELL?

The gene is the fundamental unit which determines every morphological character. It consists of DNA (deoxyribonucleic acid) and is located in the chromosomes. However, no gene acts by itself, in that, except for the character linked to sex, it is always accompanied by an analogous gene which comes from the other parent and which governs the same character. The two genes of a pair thus constituted are called alleles and the couple is said to be allelomorphic.

The number of the genetic couples constituting the hereditary patrimony of Man is not known. The estimates proposed by various authors in the past have varied from a minimum of 5000 to a maximum of 120,000. Recently (1964) Vogel has presented the question in a very original way. The data on which his deductions are based are as follows:

(1) Normal human haemoglobin is formed of 4 globin subunits, 2α and 2β to which are connected the porphyrin nuclei containing the iron. The α globin chain of haemoglobin contains 141 amino acids and the β chain of 146, which corresponds to a total molecular weight for the haemoglobin of about 67,000. This means that the α and β chains are made up of 423 and 438 pairs of nucleotides respectively.

Many other proteins consist of subunits of the same order of magnitude, that is, they have a molecular weight of about 17,000. From this it can be deduced that a functionally active gene must have a length of approximately 450 pairs of nucleotides.

(2) The weight of a haploid set of human chromosomes (the nucleus of a spermatozoon) is about $2 \cdot 72 \times 10^{-12}$g, while the nucleus of a granulocyte in the blood weighs approximately $6 \cdot 23 \times 10^{-12}$g.

Since, generally speaking, it is known that the DNA content of every cell in all species is constant, it can be accepted that the total quantity of DNA in a human haploid set would be equal to $3 \cdot 10^{-12}$g.

(3) Normally, the variants of human haemoglobin differ by the substitution of a single amino acid. Therefore it would seem that a functionally active gene can produce only one type of genetically determined polypeptide chain. This means that functionally active genes can be present only once. It is then easy to calculate that a pair of nucleotides with one adenine, one thymine, two deoxyriboses and two residual phosphates have a weight of about 1×10^{-21}g.

Now as the total haploid set of human chromosomes contains about 3×10^9 pairs of nucleotides, assuming that 450 pairs are needed for one average functionally active gene, one can deduce that in a human haploid set there will be approximately $6 \cdot 7 \times 10^6$ functionally active genes. This number is extremely high.

On the other hand, on the basis of other deductions derived from the structure of the chromosomes of *Drosophila* and the number of genes they contain, Vogel reached the conclusion that the number of genes in the human haploid complement could certainly be not less than $6 \cdot 7 \times 10^4$. According to these calculations therefore the number of genes contained in a human haploid nucleus would vary between 67,000,000 and 670,000. However, even if these calculations are invalidated by the impossibility of determining exactly what a functional gene is, they give an idea of the enormous complexity of the human genotype.

If we then wish to distinguish genes which produce the same forms in all species from those genes responsible for individual or group variations, we can assume that there are about 5000 of the latter. Now since each pair of alleles can give rise to three genotypes, there will be 3^{5000} different human genotypes for each individual genetic couple. This enormous number of possibilities demonstrates that it would be virtually impossible (except in the case of monozygotic twins) to find two genetically identical individuals existing on earth. Consequently every individual represents in himself a "unicum" unlikely to be duplicated.

4. THE INDIVIDUAL AND POPULATION

Man, like almost all the other species of primates, is a gregarious animal and as such tends to live in groups which, in a non-literal sense, are called "populations". In general the individuals belonging to one population differ in some characters from those belonging to other populations on the basis of the presence or absence of various alleles or in terms of their relative frequency.

Disparate human or animal populations differ genetically from one another almost exclusively in the varying proportions of alleles which they contain. The complete loss of one of the alleles in a population has never been noted in any of the extensively studied cases but there are different frequencies ranging to almost nil.

For any population it is possible, by means of a comparatively simple calculation, to derive the relative frequencies of the alleles which produce the various phenotypes from the proportions in which the phenotypes are present. The genetic description of a population consists therefore of a list of the relative frequencies of the various alleles which exist in it.

5. NATURE AND DESCRIPTION OF GENETIC VARIABILITY IN HUMAN POPULATIONS

If, for example, P and Q are two genes present in the nuclei of both members of a parental couple in the heterozygous state, both the parents will during gametogenesis produce $0.5P + 0.5Q$, so that the end result of their mating in terms of probability will be:

$$(0.5P + 0.5Q)^2 = 0.25PP + 0.50PQ + 0.25QQ.$$

If however, one parent is heterozygous and one homozygous, their offspring will receive:

$$(0.5P + 0.5Q) \times 1P = 0.5PP + 0.5PQ.$$

There is, therefore, an obvious relationship between parental genotypes, the gametes produced by them and the frequency of types of offspring. Consequently, for the study of the mating of different genotypes one can substitute that of the combination of the corresponding gametes, keeping in mind the relative frequency with which the various gametes will be produced in the entire group.

In fact, if we suppose that the factor P contributes to produce the genotypes of a group in the proportion of 40% and the factor Q contributes the residue of 60%, the gametes which will be produced by the entire population, whether male or female (except for sex-linked characters) will be found in the same proportion, and the frequency of the genotypes of the offspring can be found by means of the following calculation:

$$(0{\cdot}4P + 0{\cdot}6Q) \times (0{\cdot}4P + 0{\cdot}6Q) = (0{\cdot}4P + 0{\cdot}6Q)^2 = 0{\cdot}16PP + 0{\cdot}48PQ + 0{\cdot}36QQ.$$

In its turn the next generation will produce gametes P and Q in the same proportions and will produce the initial condition.

The formula $(0{\cdot}4P + 0{\cdot}6Q)^2$ is therefore suitable to represent synthetically the distribution of the genotypes in the population both at the time of the study and in successive generations, provided that the combinations among the different gametes occur by chance and that there are no external intrusions.

If the population constitutes a well-defined systematic unit, such as a race or even a species, the formula may be considered as the racial or specific allelotype.

Sometimes, however, the characters (as, for example, the blood groups ABO) are determined by multiple alleles. In this case, the genes which combine in the single allelomorphic couple are of three types (P, Q, R) and therefore the gametes produced in the aggregate of the population will be of three kinds, each of which will be present in the mass of gametes of the population with a particular frequency. The allelotype relative to that character assumes the trinomial form:

$$(pP + qQ + rR)^2$$

where p, q and r are the relative frequencies of each of the characters. The type and frequency of all the genotypes in the population can be obtained from the development of the trinomial.

6. GENETIC EQUILIBRIUM AND THE HARDY-WEINBERG LAW

From the genetic point of view, an allelotype represents a population as long as all types of mating happen without any selection and with the frequency of the genes in play remaining constant. A population in which the genetic frequencies for a determinate character can be reduced to a binomial or polynomial distribution (even if asymmetrical) is said to be in a state of "genetic equilibrium".

Let us suppose, for example, that an isolated territory is populated by an equal number of dominant TT homozygotes and of recessive tt homozygotes for sensitivity to phenylthiocarbamide, and that members of this population mate by chance. The possible matings, as far as this character is concerned, will be: $TT \times TT$, $TT \times tt$ and $tt \times tt$. If the original proportion of TT to tt is 50%, each is then equal to 0·5 of the total population considered as 1, and the gametes produced by these individuals will also be in the same proportions. Therefore, if they mate by chance, we will have:

		Eggs	
		0·5T	0·5t
Sperm	0·5T	0·25TT	0·25Tt
	0·5t	0·25Tt	0·25tt

That is to say: $0\cdot25TT + 0\cdot50Tt + 0\cdot25tt$. In the successive generation the proportions of the gametes produced will be the same as in the preceding generation since the frequency of the alleles T and t of the genes will be:

$T = 0\cdot25$ (from the homozygotes TT) $+ 0\cdot25$ (from the hetero-zygotes Tt) $= 0\cdot5$

$t = 0\cdot25$ (from the homozygotes tt) $+ 0\cdot25$ (from the hetero-zygotes Tt) $= 0\cdot5$

and so on in successive generations. It is possible to generalize that if q is the fraction of the gamete carrying the gene t, the distribution of the genotypes in the successive generation will be:

		Eggs	
		$(1-q)T$	qt
Sperm	$(1-q)T$	$(1-q)^2TT$	$q(1-q)Tt$
	qt	$q(1-q)Tt$	q^2tt

that is: $\qquad (1-q)^2\, TT : 2q(1-q)Tt : q^2tt$

In a case of simple allelomorphism, designating the genetic frequency of the allele t by q, and that of the allele T by p, yields the statement

$$p + q = 1, \text{ or } p = 1 - q.$$

The distribution of the genotypes in the generation under consideration may then be described by the following formula:

$$p^2TT : 2pqTt : q^2tt,$$

the terms of which correspond to the development of the binomial

$$(pT + qt)^2.$$

The expression $(pT + qt)^2 = p^2TT + 2pqTt + q^2tt$ is known as the Hardy-Weinberg law, and the genotypic frequencies for any allelomorphic character in a population reproducing by random mating can be derived from it. In order to establish a condition of equilibrium in a case of this sort, the sum of all the frequencies considered must be equal to one:

$$(p + q)^2 = p^2 + 2pq + q^2 = 1.$$

The Hardy-Weinberg law makes it possible to determine whether or not a population is in genetic equilibrium.

From the foregoing it is also obvious that it is possible to calculate the frequency of a recessive gene in a population when one knows the frequency of its phenotype. In fact, the frequency of an allele in a population may be derived by the calculation of the square root of the frequency of the homozygous phenotype. Moreover, when the frequency of the recessive allele of a gene is known, it is easy to calculate the frequency of the dominant one and of the heterozygotes by means of the Hardy-Weinberg formula previously considered.

We shall now consider a population in which the genotypes are not distributed in a perfectly binomial way: for example the genotypic population $0.3PP + 0.4PQ + 0.3QQ$.

Since $\sqrt{0.3} + \sqrt{0.3} = 1.096$ and thus is not equal to 1, the genotypes cannot be considered as a chance combination of two genes and therefore the population is not in genetic equilibrium.

The gametes produced by this population will be of the following types and frequencies:

$$0.3P + 0.2P + 0.2Q + 0.3Q = 0.5P + 0.5Q$$

and therefore, providing that panmixis is still the rule, the distribution of the genotypes in the successive generation will be:

$$(0.5P + 0.5Q)^2 = 0.25PP + 0.50PQ + 0.25QQ$$

which is a different distribution from the preceding one. From this moment, however, the population is once more in equilibrium. In fact the square roots of the frequencies of the homozygous genotypes, which correspond to the frequencies of the genes, have a sum of one:

$$\sqrt{0{\cdot}25} + \sqrt{0{\cdot}25} = 0{\cdot}5 + 0{\cdot}5 = 1.$$

It is thus possible to conclude that "a population in which only pan-mixis occurs, tends to conserve the state of genetic equilibrium or to recover it when it is upset by some transitory causative agent".

In practice, equilibrium will not be restored in the immediately succes-sive generation after the disturbing factor has ceased to operate. A varying length of time, which can be expressed as a number of generations and which will bear a direct relationship to the strength of the obstacles which interfere with the free circulation of genes (ethological, ecological or geo-graphical barriers and so on), will be needed.

Thus disturbances of the genetic equilibrium almost always end in the establishment of a new form of equilibium which can lead to a new allelo-type. In this way it is possible to bring about a transformation in the patri-mony of the hereditary material of a population which can lead to its profound alteration. If the differences are advantageous, the new population will survive and may even overpower the population from which it origina-ted, eventually constituting, as we shall see, a new subspecies or even a new species depending on the type and number of genes substituted.

7. Possible Causes of Variation in the Allelotype of a Population

The allelotype of a population can be altered by various factors among which the most important are mutation, selection, genetic drift in small fractions of a population, and differential migration.

Mutations can affect either the number or the structure of the chromo-somes or of the single hereditary units. The most interesting mutations at the population level are genic mutations, in that they introduce something qualitatively new in the patrimony of hereditary material of a population, and manifest themselves as variations of a discontinous nature. They may arise spontaneously or may be induced by physical agents (ionizing radiation for example) or by chemical agents. The fact that the frequency of mutations is different for different loci is of great interest. There are tables which show an extreme variability in the frequency of the appearance of mutations in man. On the whole, however, they can be calculated to occur under natural conditions with an average frequency of 1:100,000.

In general, the vitality of a mutant is less than that of the original gene, probably because, during evolution natural selection has already fixed upon those mutations which have appeared most frequently and have proved to be most advantageous to the vitality of the species. Still, if a gene presents some advantage in a particular environment, selection will favour it, and it may establish itself sufficiently to influence the general allelotype of the population. In this regard we shall later see how the pathological gene for thalassaemia is advantageously established in some human populations of the Mediterranean area.

The differential survival of genes from one generation to another is determined by external factors acting to select the most suitable. In "artificial selection" (the breeding of domestic animals) differential survival or the reproduction of the carriers of the different genotypes is controlled by human choice; in "natural selection" these are for the most part dependent upon factors in the external environment.

Physical environment (climatic and geographical) is among the natural agents which most influence the natural selection of our species. If, for example, we consider what is known as the "pigmentation complex" (skin, eye and hair colour) we see that those populations having the darkest pigmentation occupy the equatorial regions. The factor which in this case seems most closely correlated with this character is solar radiation. Connections such as this between environmental factors (solar radiation) and the physical characteristics of the equatorial populations are not accidental but, at least from some points of view (as we shall see in Chapter 3), advantageous.

In populations made up of a limited number of individuals the frequency of the genes is subject to genetic drift from one generation to another.

From a statistical point of view genetic drift can be defined as "an accidental increase of the alleles" of a particular type whereas from the dynamic point of view the phenomenon can be exemplified as follows.

If we imagine a European type population with 15% of the individuals having the gene B, and we take from it a small sample, for example four men and four women, it is highly probable that the sixteen genes in the sample would all be of type O or type A, or in any case that none of the eight individuals would be of groups B or AB. If these eight individuals should by chance be transported to a desert island their descendants would all be phenotypically A or O, and genotype B would be lost.

Isolation leads to the establishment of a new equilibrium corresponding to an allelotype with diverse frequencies of genes and, therefore, to a population different from that preceding it. Later on we shall examine some concrete cases.

However, the phenomenon of drift presupposes the isolation and the small size of the isolated groups. For this reason, it may be thought that in the human species drift must have occurred more frequently in the past when humanity was distributed in comparatively small groups separated by large distances.

Differential *migration* of part of a population to a place where it can mix genetically (hybridize) with representatives of other populations differing in the frequency and the quantity of particular genes may increase or diminish the frequency of some of these genes. In the case of man, this phenomenon is called "miscegenation" and, in the case of a great range of mixing between two different populations, "metamorphism".

In human populations a certain type of selection is also produced by non random mating. This selection may be due to endogamy or to selection for similar phenotypes or to other preferences. These events can produce a variation in the allelotype of a population. However, the direction of this variation is hard to define because in the same generation or in the following generations a reverse effect can occur such as exogamy, or marriage by preference between dissimilar phenotypes.

8. The Physical Basis of Heredity: the Chromosomes

Every cell in our bodies, like those of every other living being contains a small body which is basophilic and largely made up of deoxyribonucleic acid (DNA): the nucleus. In the interkinetic phase the nucleus appears almost homogeneous, but when the cell divides, filamentary structures appear within it called the chromosomes. These represent the character of the species at the cellular level. Each animal or vegetable species is characterized by a fixed number of chromosomes and by their typical morphology. The number of chromosomes is constant to the point that even the supplementary presence of one of the smallest (the twenty-first) in the cells of a human individual has as a consequence Down's syndrome (Fig. 2.3).

The chromosomes are most clearly visible during the metaphase of mitotic division, and appear to be constituted of two filaments (chromatids) joined together in a region that is characteristic for each chromosome (the centromere). At this point each chromosome is already self-duplicated and each of the chromatids has been drawn into one of the two daughter cells.

By means of this mechanism of self-duplication the complete body of information which the individual receives at the moment of conception from his parents is perpetuated unaltered for generations and generations of cells.

In order to maintain the absolutely constant quantity of information transmitted from one generation to another, it is necessary that the number

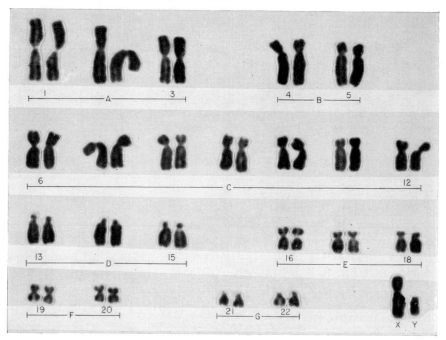

FIG. 2.3. The human caryotype ordered according to the Denver statement. The presence of a supernumerary chromosome in Group G is the cause of Down's syndrome.

of chromosomes be reduced by half so that in the zygote a set equivalent to the original can again be assembled. This reduction takes place during the maturation of the gametes.

9. CONTINUITY OF THE SEX CELLS

The sex cells, in contrast to the somatic cells, have the property of transmitting their chromosomes in perpetuity from one individual to another in successive generations. The somatic cells of an individual are destined to die at the death of the individual, the sex cells on the contrary are potentially immortal. Only the latter therefore establish a continuity from one individual to another with the succession of generations and only the variations which occur in the chromosomes of the germ cells have any importance in producing variations in the hereditary complex of a species.

In sexual organisms, in the very first divisions of the zygote, groups of cells are differentiated which, by localizing themselves in particular regions of the embryo, determine the development of the gonads. In the gonads these cells continue to multiply and to differentiate into spermatogonia in the case of the male gonad, or into oogonia in that of the female gonad.

During the maturation of the spermatozoa and the ova, the chromosomes travel by the same paths, although there are differences in the duration of the intermediate phases and the regulation of the number of gametes. When a spermatogonium in the testes ceases to multiply by mitosis and begins to enlarge, it is a sign that it is about to initiate meiotic division. The cell found in this state is called a primary spermatocyte and each primary spermatocyte is destined to form four spermatozoa. The result of meiotic division is the reduction of the number of chromosomes by one half (Fig. 2.4). The haploid number of chromosomes contained in the spermatozoon is the essential contribution of the male to the hereditary complex of the offspring.

Oogenesis differs from spermatogenesis principally in the behaviour of the cytoplasm during meiotic division. In order to accumulate nutritive substance the primary oocyte is much larger than the primary spermatocyte.

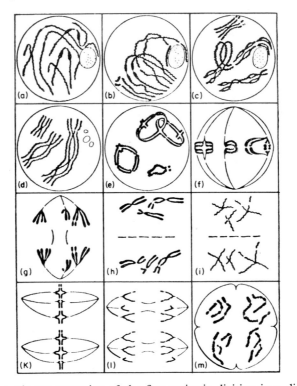

FIG. 2.4 Schematic representation of the first meioetic division in a diploid cell: leptotene; (b) zipotene; (c) pakitene; (d) diplotene; (e) diakinesis; (f) metaphase I; (g) anaphase I; (h) telophase I; (i) interkinesis; (k) metaphase II; (l) anaphase II; (m) telophase II (from Riger, 1968).

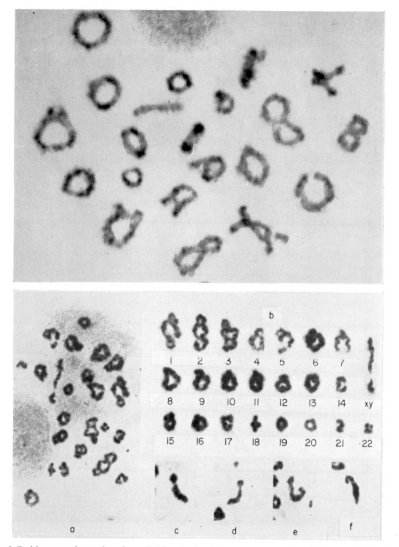

FIG. 2.5 Above: plate showing diakinesis in *Macaca nemestrina*. Below: diakinesis in *Homo sapiens*; in (b) the paired chromosomes are arranged in decreasing size and c, d, e and f, show the paired XY in other plates.

The first two meiotic divisions are extremely irregular regarding the quantity of cytoplasm they contain. The cell which will become an active ovum receives virtually all the nutritive substance accumulated in the cytoplasm, while the others (polar bodies) receive an insignificant portion. In this way, if the egg cell should be fertilized, it can develop into an embryo while the

polar bodies cannot. The behaviour of the chromosomes during the maturation of the ovum is exactly the same as that seen in the sperm except that it
takes place at different times.

10. Chance Distribution of Chromosomes and Genes During Meiosis

The fact that only one division of the chromosomes occurs during the two
meiotic divisions inevitably results in a reduction of the number of chromosomes in the gametes by one half with respect to the other cells of the
organism. In one phase of the division the two homologues separate and
consequently each mature gamete possesses only one element of each pair.
Which of the two homologous chromosomes goes to a particular gamete is
decided by chance, so that the various possibilities are realized with equal
frequency. This is, therefore, a factor which contributes notably in keeping
the association of hereditary characters random within the limits of a species.

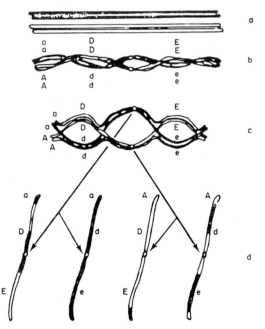

Fig. 2.6 Diagrammatic representation of the recombination at meiosis. Each unit is
constituted of two filaments, one of maternal and the other of paternal origin. Paired at the
beginning, (a), they are coiled further on, (b); and later exchange some parts (c, d). The
letters Aa, Dd, Ee, indicate genes and the figure shows them as they disperse after the
second meiotic division into filaments which become part of a germ cell. This happens
independently for each of the 23 pairs, and every germ cell receives a complex mixture of
genes with paternal parts and parts of maternal origin.

However, another mechanism operates to render still more haphazard the combination of genes in offspring. This mechanism consists in the exchange of portions of chromosomes between homologues during the first meiotic division and is called "crossing-over". Crossing-over can be represented schematically as follows: During the stage of synapsis the two homologues attract each other and come to lie side by side. While they are thus coupled their chromatids are reduplicated with the result that the entire unit consists of four filaments. Two of these are of maternal origin and two are paternal. Now, during the process of duplication or immediately after, it often happens that two of these filaments exchange segments. That is, for some yet unknown reason a maternal and a paternal filament break simultaneously at the same point. The broken pieces weld themselves reciprocally in such a way that two new chromatids are formed, each with one piece of paternal origin and one piece of maternal origin (Fig. 2.6). This particular stage is recognizable under the microscope and is called diplotene. The point at which the chromatids cross over is called the point of chiasma. Each of the four chromatids thus formed has the same chance as the others of entering into a gamete and so in forming part of the genetic make-up of an individual. Now since each chromatid pair may break and join up by chance at any point in its length, the possibility of an offspring inheriting many of the genes found associated together in one of the parents is made still less likely.

11. How Chromosomes May Change in Number and Form

The systematic and experimental study of the variations which the chromosomes in plants and animals may undergo in number and morphology has made it possible to identify a number of mechanisms capable of altering the karyotypes of individuals or of groups of individuals to the point of isolating them reproductively from other representatives of their species. There are diverse means by which the karyotype of a species may come to differ in number or morphology. Among numerical variations polyploidy (duplication of the whole chromosome set) has certainly played a very important role in the evolution of vegetal species and of many lower animals. In the case of the mammals, and therefore of the primates, this mechanism may be disregarded because of the difficulty interposed by the presence of well-differentiated sex chromosomes.

Single chromosomes may be increased in number by meiotic or mitotic "nondisjunction". Phenomena of this type have been noticed not only in plants and animals but also in a number of chromosomes of the human complement and are called trisomies. The most noted of these is that

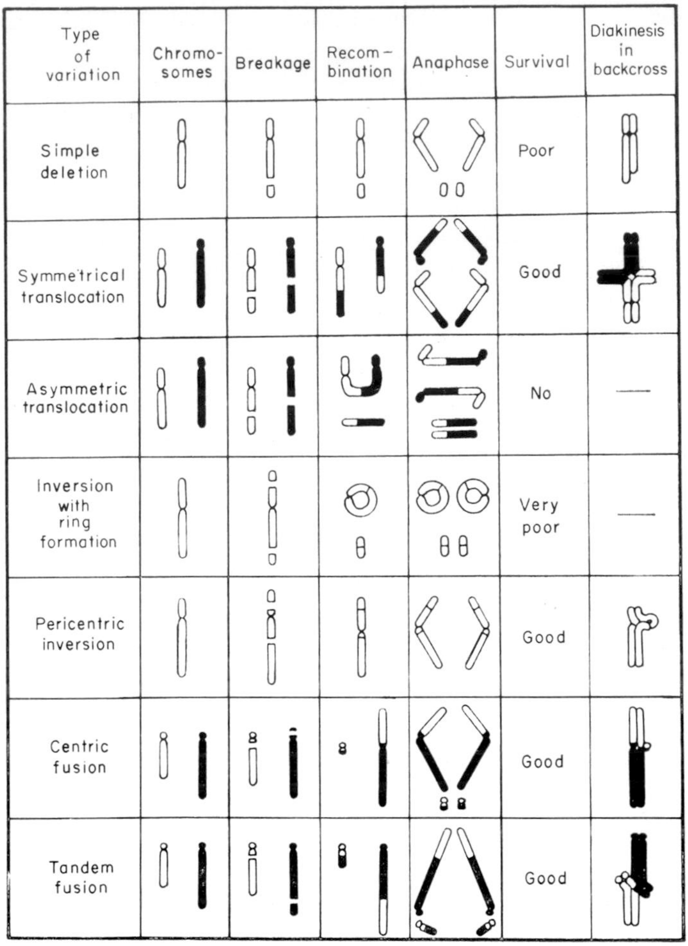

Type of variation	Chromo-somes	Breakage	Recom-bination	Anaphase	Survival	Diakinesis in backcross
Simple deletion					Poor	
Symmetrical translocation					Good	
Asymmetric translocation					No	—
Inversion with ring formation					Very poor	—
Pericentric inversion					Good	
Centric fusion					Good	
Tandem fusion					Good	

FIG. 2.7 Graphical representation of the most important structural variations of chromosomes and estimation of their possibilities of survival in somatic cells and at meiosis.

characteristic of Down's syndrome (chromosome 21). Generally they represent deleterious conditions, but there is no reason to believe that this is always so. Among plants, for example, perfectly viable trisomies are known for each of the 24 chromosomes of *Datura stramonium*.

Another theoretical means of increasing the number of chromosomes is the possibility of some of them reduplicating asynchronously in a cell. Such a duplication, if it occurred during the first mitotic division of the germinal line, could have some importance in the evolution of a species, but to date there is little evidence that this happens.

Another mechanism which can lead to an increase in the number of chromosomes, without any variation in the genetic material, is the transverse misdivision of the centromeres. In this case the region of the centromere divides transversely rather than longitudinally during mitosis, so that the chromosome divides into two chromosomes, each with a functioning centromere.

As well as increasing in number, the chromosomes can also be reduced. The mechanism which leads to a reduction in the number of chromosomes without causing grave damage to the genome is that of centric fusion. This occurs between chromosomes having the centromere in the terminal position (acrocentric) and consists in the union of two of these in this region. In other words, the phenomenon is the opposite of "misdivision" of the centromere.

The reduction of the number of chromosomes by means of centric fusion has been demonstrated in the most diverse organisms. The number of chromosomes undoubtedly has some evolutionary importance, and yet it is difficult to determine the reason. Probably it corresponds to particular requisites of an ecological nature. In fact it is possible that a low chromosome number could represent an advantage for a very specialized animal. Such a condition would create blocks of genes which would tend to be grouped together during meiosis and thus reduce the possibility of harmful deviations in the descendants.

Chromosomes may vary not only in number but also in their morphology. There are a number of ways in which these variations may come into being. All, however, involve at least one break in the continuity of the chromosome (Fig. 2.7). Breakages occur spontaneously with a rather low frequency but they occur in all organisms. The majority of changes which are important in the chromosomal mechanisms of evolution require at least two contemporary breakages in the same cell. Very often the broken pieces rejoin. At other times, however, the sections reunite in a different way giving rise to morphologically new chromosomes. The frequency with which these rearrangements are produced is low, but they do occur.

The possibility of the survival of such morphological mutations of chromosomes is further lessened by other factors which limit them either at the level of cell multiplication (e.g. bicentric chromosomes) or at that of the reproduction of the individuals carrying the variation.

If the chromosome breaks at only one point, the part lacking a centromere may re-attach itself to the same or to another chromosome. If this does not happen, the portion without a centromere, not being able to be drawn to one of the poles, is lost in the daughter cells and what is called a "deletion", that is, the loss of a portion of the genome, consequently results. If this loss of a piece of a chromosome persists in a population, it may become a homozygote and represent a variation of the karyotype. Yet a deletion, especially of a large portion of a chromosome, must be considered a rare phenomenon, in that information is lost which must certainly be useful for the survival of the individual and of the species.

If the breakage involves more than one chromosome an exchange of portions of the chromosomes may result. This type of exchange is called "reciprocal translocation". The reattachment may take place in two ways: either with the acentric fragment of the one and that containing the centromere of the other, or with the two acentric fragments together, on the one hand and, both of those having a centromere on the other.

First, two chromosomes result that are different in aspect yet perfectly viable for the cell. They are able to survive meiosis without difficulty. In the second case, on the contrary, the result is one acentric chromosome and one bicentric. Both are unable to distribute the chromosomal material during mitosis in the cell containing them and are, therefore, lethal. (The mechanism of centric fusion is nothing other than an extreme form of unequal translocation of two chromosomes with terminal centromeres.) If more than one breakage is present in a single chromosome the fragments may reunite so that the original form is changed. Also in this case it is possible to have two systems re-joining different from the prototype: (1) the free extremity of the portion having the centromere can give rise to a new chromosome and two acentric fragments; (2) the portion lying between the breaks may turn through 180° and join up again. In the first case there will be a loss of chromosomal material that is neither advantageous to the cell nor to the individual, and the formation of a ring-shaped chromosome which in mammals does not seem to be successful (it has been found only occasionally in tumour cells).

In the second case the rotated fragment may involve the centromere (pericentric inversion) or not involve it (paracentric). The morphological variation which results will manifest itself only if it involves a fragment with the centromere in an asymmetric position. The inversion is, however, perfectly viable for the cell and it can survive meiosis without serious inconvenience.

Evidently several mechanisms may intervene at the same time or one mechanism may have several effects in the differentiation of the karyotype of a species, and thus it is not always easy to distinguish the action of one of these.

GENERAL REFERENCES

Anfinsen, C. B. (1959). "The Molecular Basis of Evolution", J. Wiley & Sons, New York.

Buettner-Janusch, J. (1966). "Origins of Man—Physical Anthropology", J. Wiley & Sons, New York, London, Sydney.

Chiarelli, B. (1968). From the karyotype of the Apes to the human karyotype. *S. Afr. J. Sci.* **64** (2), 72–80.

Dobzhansky, T. (1955). "Evolution, Genetics and Man", J. Wiley & Sons, New York and London.

Florkin, M. (1966). "Aspects moleculaires de l'adaptation et de la phylogenie", Mason & Cie, Paris.

Jukes, T. H. (1966). "Molecules and Evolution", Columbia University Press, New York and London.

Lasker, G. W. (1961). "The Evolution of Man. A Brief Introduction to Physical Anthropology", Holt, Rinehart & Winston, New York and London.

Mather, K. (1964). "Human Diversity", Oliver & Boyd, Edinburgh and London.

Neel, J. W. and Schull, W. J. (1954). "Human Heredity", University of Chicago Press, Chicago.

Parenti, R. (1954–1955). "Biologia delle razze unmane", Libreria Goliardica, Pisa.

Rieger, R., Michaelis, A. and Green, M. M. (1968). "A Glossary of Genetics and Cytogenetics. Classical and Molecular", Springer-Verlag, Berlin, Heidelberg, New York.

Tax, S. and Callender, C. (eds) (1960). "Evolution after Darwin", University of Chicago Press, Chicago and London.

Turpin, R. and Lejeune, J. (1965). "Les Chromosomes Humains (Caryotype Normal et Variations Pathologiques)", Gauthier-Villars, Paris.

Vogel, F. (1964). A preliminary estimate of the number of human genes. *Nature, Lond.* **201.**, 847.

3

Action of the Environment on the Evolution of Populations and of Species

1. GENERAL

The phenotype, the external aspect of an individual, is simply the result of environmental modification on the genotype, the genetic make-up of the individual. Thus environmental action is responsible for character variability in individuals, such as in monozygous twins, having an identical genotype.

Although environmental action is a very important factor in individual variation, it displays its most significant and most obvious effects at the population level. The isolated individual, even though he constitutes the fundamental unit, has little relevance in evolution. Individuals belonging to one population in general differ from those of another population in a number of characters. These differences are almost always the result of the mechanisms of selection and adaptation of a determinate population to a particular environment.

We must ask, however, how the environment acts on an entire population, most particularly on a population such as the primates and even more to the point, on man.

First of all, we must define what we mean by environmental action. All the climatic factors of the biosphere are potential selective agents which create differences in the population by means of genetic adaptation. Quality and quantity of food are also important selective agents. Abundance of food favours the survival of offspring; varied and nutritionally complete foods prevent the development of deficiency diseases such as avitaminoses or growth defects. Epidemic diseases also act as selective factors, favouring those individuals who present the greatest immunity to the pathogenic agent. The natural environment which surrounds us is thus full of forces which mould populations and differentiate between them.

When a population is especially well adapted to a climatic zone or immune to particular diseases, there have always been selective factors which have caused the differentiation. However, there are many environmental factors, for the most part little known, and it is, therefore, not always easy to evaluate the differentiating potentiality of a particular environment. How, for example, are we to interpret the evolutionary significance of the Mongolian fold ? What is the reason for the high frequency of Duffy's genes among the Bantu ? Why do some pathological conditions present a peculiar geographical distribution ?

In the following we shall consider some of the most striking cases of environmental action as a selective agent.

2. Climatic, Nutritional and Immunological Differences

Of all the elements of the natural environment which surrounds beings such as man or the other primates, none has been as important as climate. Climate must be understood as the totality of the factors which make up the atmospheric environment, with their periodic variations of temperature, pressure and humidity, and including occasional and sometimes catastrophic events such as hurricanes. The most important factors in climatic differentiation are temperature, humidity, the intensity of solar radiation and altitude above sea level.

In the biosphere the temperature oscillates between a maximum of $+50°C$ in the shade to a minimum of $-30°C$, that is with a difference of about $80°C$, but these are extreme temperatures which are found only in certain regions. Each part of the earth displays different characteristics of temperature with more or less uniform seasonal rhythms.

In certain zones, for example in desert areas, the changes of temperature are extremely rapid between day and night, with variations of more than $30°C$. within a few hours. The inhabitants of the Australian desert, of the Kalahari or of some desert regions of the southwest of the United States may equally well burn their feet at noon and freeze them during the night.

Clothing, houses, sun shades, central heating and air conditioning mitigate the effects of variations of temperature on their bodies for many men, but this has not been achieved for all. It is also true that these artificial means of protection have been introduced only recently.

When we read in the newspapers of cases of death from exposure to the sun or to cold, we may begin to realize how brutal natural selection based on temperature change must have been hundreds or thousands of years ago.

The characteristics of a climate are determined not only by temperature, however, but also by the quantity of humidity present in the air, the amount of solar radiation, altitude above sea level and other factors. All of these display a wide range of variability which puts to the test the ability of individuals and, therefore, of populations to survive.

In America and in Europe there are, at present, unlimited supplies of food available. At little cost we are able to feed ourselves to satiation, ingesting daily as many as 6000 calories, while our average daily requirements are about 3000 to 3500. Overeating is one of the most frequent causes of death, in some of these regions. In other parts of the world, on the other hand, 1500 calories per day constitute an exceptional diet; for much of the population of southern Asia 800 calories daily represents the normal condition. Thus, as for temperature, the extremes of the quantity of food ingested are very far apart. Nevertheless death from starvation is far more common than death from heat or cold.

Other than climatic factors and scarcity or excess of food, there are numerous infectious diseases (cholera, typhus, malaria, diphtheria, tuberculosis, and dysentery) which exterminate, and have exterminated, populations of millions of individuals. In various parts of the world malaria, smallpox, amoebic dysentery, and poisonous snake-bites are common causes of death. Even today in many depressed areas it is still rare for one child in three to succeed in reaching adulthood.

The differentiating action at the level of populations, of many diseases or complexes of infectious diseases is still fairly obscure.

3. Stature, Body-size and Natural Selection

Among living human populations variations in stature and body-weight are remarkable. The average height of males can vary from 185cm in one population to less than 152cm in another. The body weight of a thin European is approximately 61kg while in other ethnic groups individuals of medium thinness may weigh in the neighbourhood of 47kg.

Large body dimensions may confer advantages; a big man commands respect, and in the past must have been favoured in hand-to-hand combat and in hunting large animals. Moreover, a strong man can occupy a larger territory; he is generally more resistant; he can undertake hunting expeditions which require a greater use of force and so carry home a larger quantity of meat. On the other hand, great height and corpulence can also be disadvantageous. Large body-size requires many calories, both for maintenance and particularly during growth. A child of large body-size by genetic constitution is under a great disadvantage when food is scarce. In times of famine small individuals have a better probability of survival.

There is certainly a relationship, even if it is not always very clear, between relative body dimensions and geographical distribution in relation to temperature. This results partly from physiological adaptation and partly from selection.

The quantity of adipose tissue is particularly significant. Its distribution differs according to sex, but also according to the climate of the environment. The supra-zygomatic and circumorbital fat of the Eskimo, even when this, not only contributes to the well-known round appearance of the face, but also provides a protective mask against the rigours of polar cold. The fat stored in the glutei of the Bushmen and Hottentots constitutes an obvious solution to the climatic and energy-reserve problem. It forms a reserve of calories to use in periods of scarcity which, however, being limited to a well-defined zone, interferes little with the necessity for an easy dispersal of calories to offset the excessive equatorial heat.

4. ADAPTATION TO CLIMATIC DIFFERENCES

Among the extremes of climate in which man has lived and is forced to live now intense cold is the most selective. In certain arctic climates, without adequate equipment, a man will perish in a few minutes. Without considering the possibility of using modern equipment or the construction of special shelters (igloos), the local populations present specific biological adaptations, among which the most obvious is the reduction of body surface, which consists of the reduction of the limbs, principally the legs of the Eskimo (Fig. 3.1). At very low temperatures some populations are also able to raise their basal metabolism.

In certain regions in which the changes of temperature between night and day are important, the indigenous populations often have the ability to raise their surface temperature during the night by increasing superficial circulation.

A characteristic example of adaptation to cold is that described by the Soviet scholar who tried to adapt monkeys, such as Cercopithecines and Macaques, which live in temperate and tropical climates in comparison to the climate of Moscow. These animals reacted to the severity of the winter climate of Moscow by eating more avidly and increasing their subcutaneous fat; the length and thickness of their fur also increased notably. Notwith-

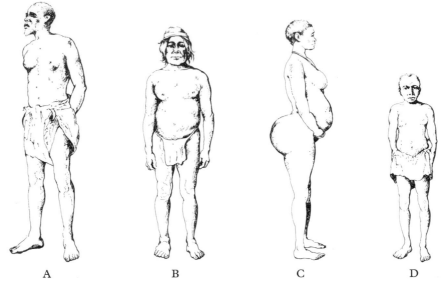

FIG. 3.1 Examples of human environmental adaption: the Nylotic or Sudanese Negro (A), the Eskimo (B) the Hottentot (C) and the pygmy (D).

standing the first difficulties, their survival in Moscow has been excellent, so much so that they have also produced offspring.

A typical example of adaptation to variations in altitude is the increase in red blood cells which is seen in all members of our species who habitually live at high altitudes.

Life in desert regions demands a remarkable tolerance to diurnal heat and nocturnal cold without the heat balance being disturbed: that is, it requires the capacity to discharge calories by transpiration in such a way as to conserve the largest possible quantity of water, which in such areas is a precious commodity. The various types of desert mammals display different adaptations to an arid climate: nocturnal habits, leanness, and reduced body stature. Among human populations, the individuals most adapted to desert climates are the Bushmen. In fact, the pigmentation of the skin of men who live in desert climates (hot-dry) must not be too light, for protection against the action of ultraviolet rays but at the same time must not be too dark, in order to facilitate the dispersion of calories. The Bushmen, with their yellowish-brown pigmentation and the peculiar distribution of their adipose tissue, represent an example of compromise between different aims of selection in competition.

Another type of specialization found among desert rodents is an increased ability to concentrate urine. If man were able to follow the example of the desert rat, he would be able to save much water every day. At this moment definite evidence for or against such an increase in concentration of urine in human populations adapted over a long period to desert regions does not exist, but there are a number of indirect indications that make it probable that this may come about.

A hot-humid environment is also injurious to man because it depresses physical activity. If the air is loaded with humidity and the temperature is elevated, the body reacts by increasing perspiration. In the tropics, and in general in hot, humid climates, serious disorders arise from the excessive loss of salt caused by sweating. Salt is an important dietary necessity, and is particularly lacking in many tropical regions. If body temperature rises too much in a condition of salt loss, muscular spasms and cramps result, and a chronic condition of deficiency can lead to death from cardiac collapse.

These are some of the most obvious cases of environmental action in determining morphological and physiological modifications in whole populations. But, at the moment we are unable to say how these factors operate at the level of single genes, causing their frequency to vary within a particular population or altering their manifestation.

5. UNUSUAL FREQUENCIES OF ABNORMAL GENES IN HUMAN POPULATIONS

An example of how the environment may exert an influence on some genes is provided by the peculiar frequency of genes normally considered pathogenic in malarial zones. According to the Hardy-Weinberg law the constitution of genes in a population must be in continuous equilibrium and, moreover, it seems logical to think that anomalous genes and those carrying pathological conditions will be eliminated by the action of natural selection. However, these conditions are not always respected and hereditary characters normally considered pathological may present a very elevated frequency.

The explanation of these facts should be sought in the advantage that one of these genotypes may present in particular environmental conditions now existing or existing in the recent past. We shall try to give some examples.

Humanity has had to combat malaria for generations. The countries lying between 35°N. and 20°S. in both hemispheres have been particularly affected and could be called the cradle of malaria. In Europe and in the Middle East from Roman times onward malaria has existed on both shores of the Adriatic, in Greece, in the Ionian Islands, in Crete and on the shores of the Black and Caspian Seas. Given that the malarial zones were generally densely populated in the past it is not surprising that in 1880 the highest death rate for an illness in our species should have been ascribed to malaria.

In the regions of southern Europe and eastern Africa where malaria was endemic, it was noted that some individuals were not subject to malaria, or at least they did not display the characteristic rigours, the high and recurrent fever, the enlargement of spleen and the typical intestinal symptoms. Successive attempts were made to explain these facts, yet no satisfactory trial was made to determine whether the immunity was natural or acquired, whether it was due to some transitory cause or whether there existed a primary immunity to malaria. Genetic mechanisms have recently been discovered which confer a special immunity and this constitutes one of the most interesting examples of polymorphism in man. One of these is thalassaemia.

In Italy, especially in Sicily, in Sardinia and in the region of Ferrara, thalassaemia is very common (Fig. 3.2) and displays two forms: the major (Cooley's Anaemia) and the minor. The study of genealogical trees reveals that thalassaemia is hereditary and that the parents of the individual affected by thalassaemia major are carriers of the disease and display the minor form. This affliction is inherited as a dominant mendelian gene. The heterozygous state corresponds to thalassaemia minor, the homozygous state to the major form.

From the incidence of the disease it is possible to calculate the frequency

FIG. 3.2 The distribution of thalassaemia in Italy. The number indicates the percentages.

of the gene, which ranges from almost 0 in Switzerland to 0·20 or more in certain zones of Cyprus and northeastern Italy.

For the greater part of Italy, where the epidemiological data are complete, the frequencies for thalassaemia vary from 0·2 to 0·06 in Sicily and in Corsica, to about 0·10 in coastal zones northeast of Bologna and 0·18 in the region around Ferrara.

The epidemiological data, like the genealogical, confirm the hereditary character of thalassaemia.

Research carried out on the blood of individuals affected by thalassaemia major and on heterozygous individuals (minor) has demonstrated the incapacity on the part of these individuals to produce the normal haemoglobin A of the adult. The haemoglobin of the blood of those suffering from thalassaemia major is, therefore, almost entirely of the foetal type, while in thalassaemia minor normal haemoglobin is present but in reduced quantities.

The disadvantage of the homozygotes thus consists of the inability to produce adequate quantities of haemoglobin of the adult type and the symptoms which characterize them can for the most part be referred to this deficiency. For this reason almost all the homozygous individuals, that is those affected by thalassaemia major, die very young and few reach sexual maturity, so that in every generation a certain proportion of the genes for thalassaemia are eliminated. The clinical and geographical data on thalassaemia raise two questions: the first relative to the origin of the abnormal gene and to its restriction to the Mediterranean, and the other relating to its persistence in time.

From the distribution of the gene on both sides of the Mediterranean it is thought to have had a very ancient origin; in fact it probably appeared about 5000 B.C., before the expansion of the Mediterranean populations. The absence of thalassaemia from other parts of Europe where Mediterranean populations have emigrated is, however, a difficult fact to explain.

The perpetuation of the abnormal gene is still more peculiar. Since in each generation a certain number of abnormal genes are lost through the early death of homozygous individuals, it would be expected that this pathological condition would be present at a low level, balanced only by new mutations. How then can the perpetuation of the abnormal gene and its high frequency in Crete, the region of Ferrara and Sicily, be explained?

An explanation might lie in an increase in the fertility of the heterozygotes, but this has not been proved, nor on the other hand would it explain the high incidence of this gene in some places and not in others. Moreover, the mutation rate for this gene has not been found to be particularly high, nor would this explain the remarkable variation in incidence of thalassaemia in Italy, Greece and elsewhere, from Iran to southern China (Fig. 3.3).

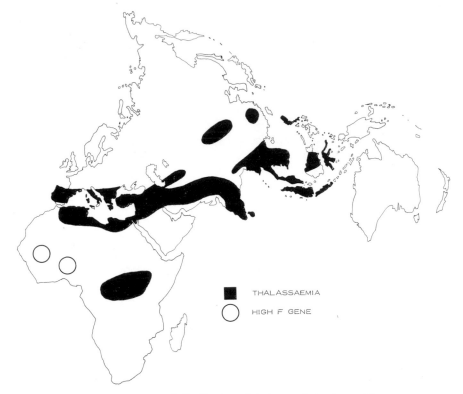

FIG. 3.3 General distribution of thalassaemia in the world.

In 1950 a group of investigators turned their attention to the correspondence between the distribution of thalassaemia and the incidence of malaria. In those parts of Europe where malaria was completely endemic, the frequency of thalassaemia was highest, while in regions having a cold climate and in the highlands the frequency approached zero. In the Sardinian populations, for example, the frequency was much higher in the low coastal regions, where malaria was endemic, than in the same type of village at higher altitudes. These data demonstrate that persons suffering from thalassaemia are in some way immune to malaria.

In a malarial zone, in fact, normal individuals are affected by malaria and there is a high degree of probability that they will die young. Thalassaemic homozygotes also die at an early age from Cooley's anaemia. The heterozygotes, on the contrary, are protected from malaria and, even if they are handicapped by a deficiency of haemoglobin, the continuation of both genes is assured through them.

This simple interpretation agrees with the data and explains the non-

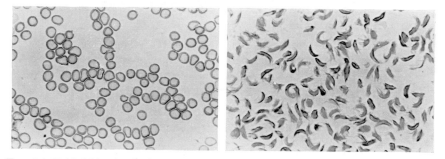

FIG. 3.4 Sickled blood cells found in patients with sickle-cell anaemia. Normal red cells in the left, sickled red cells in the right.

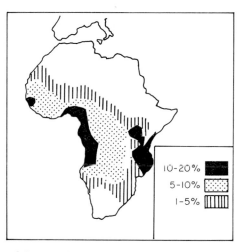

FIG. 3.5 Distribution of the gene for *Falciparum* anaemia in Africa. The high frequencies of the gene are confined to regions where tertiary malaria is a common cause of death.

uniform distribution of thalassaemia in the Mediterranean. Thalassaemia, in malarial zones, persists by means of natural selection.

Another abnormality of the blood generally diffused in Africa is sickle-cell anaemia, so-called because of the particular shape assumed by the afflicted person's red blood cells when they are immersed in a saline solution (Fig. 3.4). Like thalassaemia, the condition is hereditary and is known in two forms: one more and one less serious. The molecular nature of the disease was established in 1949 through study of the haemoglobin.

In Africa frequencies of the heterozygote vary from zero up to 40% in certain zones (Fig. 3.5).

Initially these wide variations were interpreted as being due to environmental influence, and emigrations from areas of high to areas of low frequency were postulated. However, as with thalassaemia, a more profound interpretation was necessary to explain both the perpetuation and the distribution of sickle-cell anaemia. Since sickle-cell anaemia is also usually lethal, there is a differential elimination of homozygous individuals, and it is evident that the gene for sickle-cell anaemia must present some selective advantage. In this case also the advantage consists in a better possibility of survival in a malarial environment. Once again, in fact, a strict association exists between the geographical distribution of sickle-cell anaemia and of malaria.

The advantage of selection in favour of heterozygotes has been confirmed in particular by following the destiny of children in the hyperendemic malarial zones in Africa. In these zones the normal homozygotes are stricken by malaria while still very young and either die or survive with a much reduced vitality.

Individuals homozygous for sickle-cell anaemia suffer a high rate of mortality because of the disease. The heterozygotes, on the other hand, are protected in the sense that although they may contract malaria they do not develop symptoms as severe as those of normal patients.

Sickle-cell anaemia in Africa, like thalassaemia in the Mediterranean, is an example of adaptive polymorphism in man.

Both the normal gene for haemoglobin A and the abnormal one responsible for haemoglobin S (sickle-cell anaemia) continue to exist in Africa because heterozygosity represents an advantage. The more common malaria is, the higher the frequency of the gene for sickle-cell anaemia. In areas where there is no malaria, heterozygosity carries no advantage. For this reason land reclamation and the use of insecticides, protective nets and anti-malarial products should lead to a reduction of the incidence of sickle-cell anaemia in zones where malaria has been prevalent in the past.

6. Sensitivity to Primaquine, Favism and other Hereditary Pathological Conditions

All medications should be used with caution. Primaquine is a product normally used as an anti-malarial. Some individuals have demonstrated a particular sensitivity to this drug. In these people the administration of the substance causes a haemolytic anaemia by destroying the membranes of the red blood cells.

Individuals sensitive to this medication have a reduced capacity for carrying oxygen in their blood; it is possible that for this reason a massive infection of *plasmodium malariae* cannot occur in their blood. As in the case of people with abnormal haemoglobin, the erythrocytes of the primaquine-sensitive do not give the parasites the opportunity of surviving long enough to permit them to reproduce. Thus individuals sensitive to these substances are immune to malaria.

The haemolysis is caused by the lack of the enzyme glucose-6-phosphate-dehydrogenase in the red blood cells, a deficiency which is especially common in many malarial zones (Fig. 3.6).

Another pathological condition which involves the red blood cells is favism. This is an allergic-type response to the pollen and to some substances contained in the pods of broad beans.

Sensitive individuals are stricken by haemolytic crises not only when they eat the pods or seeds of broad beans, but also if they walk across a field of these plants when they are in flower. People affected by favism are found only in Mediterranean regions (France, Spain, Italy, Armenia, Israel).

Recent research has demonstrated that this character is hereditary and is located on the X chromosome. It seems that a relationship between favism and sensitivity to Primaquine exists in the biochemical mechanism which determines the rate of agglutination in the red blood cells.

FIG. 3.6 Geographic distribution of glucose-6-phosphate-(G6PD) dehydrogenase deficiency.

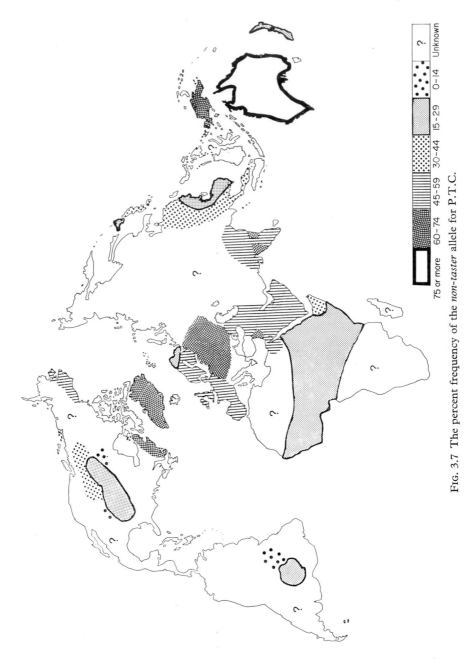

Fig. 3.7 The percent frequency of the *non-taster* allele for P.T.C.

This character is also frequent in malarial zones and, therefore, would seem to constitute an advantage against malaria.

Many other hereditary pathological conditions are known which present different frequencies in different populations. For many of these there is reason to believe that in some way they represent an advantage in the given environment.

7. SENSITIVITY TO PTC IN MAN AND IN THE PRIMATES

Another case of genic polymorphism which constitutes an excellent example of interaction between endocrine activity, taste sensitivity and food-selection, is to be found in sensitivity to phenylthiocarbamide. PTC is an organic compound which, among human beings, seems bitter to many individuals ("tasters"), while others are not aware of any taste ("non-tasters"). The character is almost certainly due to a single pair of alleles, of which the "taster" allele appears to be dominant over the "non-taster".

The frequency of the taster allele varies from 63% among Arabs to 98% among Amerinds; the same data expressed in terms of genic frequency for the recessive gene (non-taster) are respectively 37% and 20%. The geographical distribution within our species is peculiar (Fig. 3.7).

There are even populations located close together, for example the Lapps and the Scandinavians, that have very different genic frequencies, which would indicate a different history of selection for this gene. This has probably been brought about either by genetic drift or by the populations having different origins.

The primates also show a difference in gustatory sensitivity to this substance, and in these animals the character is again hereditary. A test of the threshold of sensitivity to PTC carried out on chimpanzees demonstrated a close similarity with human sensitivity (see also p. 123).

Population data have revealed remarkable differences in the frequencies of tasters and non-tasters in various groups of monkeys. In general the Cercopithecidae have a genic frequency for tasters (dominants) lying between 70% and 80%, more or less equal to that of man, while the platyrrhine monkeys differ widely from one species to another. They represent the highest and the lowest frequencies for this gene (Fig. 3.8).

The anthropoids appear most peculiar in regard to this character: the orangs have a high frequency of non-sensitivity, while the frequencies of sensitivity for the gorilla and the chimpanzee are almost the same as those for man.

The data for the platyrrhine monkeys and for the orangutan indicate the possible existence of mechanisms of genetic drift.

Taster polymorphisms to PTC seems to have existed and to have been

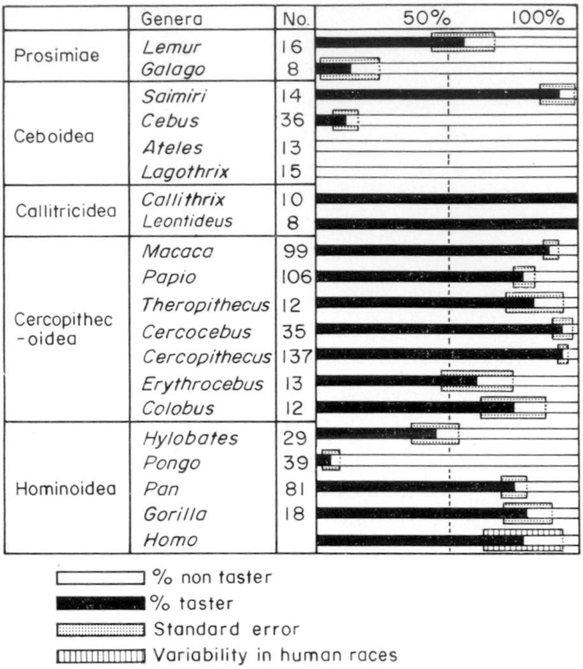

	Genera	No.	50% 100%
Prosimiae	*Lemur*	16	
	Galago	8	
Ceboidea	*Saimiri*	14	
	Cebus	36	
	Ateles	13	
	Lagothrix	15	
Callitricidea	*Callithrix*	10	
	Leontideus	8	
Cercopithec -oidea	*Macaca*	99	
	Papio	106	
	Theropithecus	12	
	Cercocebus	35	
	Cercopithecus	137	
	Erythrocebus	13	
	Colobus	12	
Hominoidea	*Hylobates*	29	
	Pongo	39	
	Pan	81	
	Gorilla	18	
	Homo		

% non taster
% taster
Standard error
Variability in human races

FIG. 3.8 Graphical representation of the percentage of taster and non-taster for different species of primates.

maintained in different species of primates for millions of years, though we cannot be absolutely certain whether it is caused by homologous genes.

To explain this fact it is possible at the moment only to formulate hypotheses. It is very probable that the tasting character in the primates may be connected, in some way, with the diet, especially with food plants such as the Brassicaceae which contain $NC = S$ substances. These substances are demonstrated to be goitrogenic, and in humans, the taster status has been established to have an effect on certain thyroid diseases.

It seems therefore that selection of a thyroid function through nutritional factors could be responsible for maintaining and modifying the taster polymorphism in the various primate species.

8. DIFFERENCES IN HUMAN PIGMENTATION, ENVIRONMENTAL ACTION AND NATURAL SELECTION

Another character, which is of great interest anthropologically and which in all probability is related to environment, is the different skin-colour found in different human populations.

The variations in the pigmentation of the skin of man have an adaptive value, and are perhaps the most striking example of the adaptation of a human being to his environment. All of us know from personal experience how even a slight tan on the most superficial part of the skin has a protective function. At the beginning of summer, before we begin to acquire a tan, even a brief exposure to the sun's rays can cause annoying discomfort. Later, with a good deep colour, we can lounge for hours under the hot sun of August without the slightest distress. A darker skin represents a natural protection against the more harmful wave lengths of the solar spectrum, the ultraviolet rays (around 2200–2800Å).

A larger quantity of melanin does not only have a protective function against the immediate damage of solar rays to the deepest layers of the skin, but this protective property also functions on other levels. For example, epithelial tumours are found more frequently among the whites of Texas than among those of regions more to the north of the United States and there are also significant differences between the incidence of this type of tumour in negroes and in white-skinned people, even when they are doing the same type of work.

The advantages which skin-pigmentation offers in climates rich in sunlight may change to disadvantages in darker climates. The transformation of ergosterol into vitamin D in the skin, essential for the growth of the bones, is catalysed by the energy coming from solar radiation. A heavily pigmented skin can represent a disadvantage where the vitamin D administered by mouth is insufficient, since it leads to complications at the time of birth and to infantile rickets. But an excess of vitamin D can be as harmful as a deficiency. A prolonged excess of vitamin D leads to the formation of deposits of calcium in the arteries and especially in the aorta, with consequent damage to renal function through the formation of calculi.

The average daily requirement of vitamin D for an adolescent is 400 international units, which are easily supplied in an ordinary diet and can be supplemented by the use of cod liver oil. Doses higher than 20,000 units daily in adolescence are harmful. In northern regions the majority of white adults manufacture all the vitamin D they need in the subcutaneous layer of the skin of their hands and faces during occasional exposures to the sun. An adult would become ill if he were receiving more than 100,000 units per day.

In the tropics a white-skinned person exposing his entire body to the sun for 6 hours daily could produce more than 800,000 units, and this, over a long period, could be very serious.

The necessity for protection on the one hand, and the interference with the synthesis of vitamin D on the other, satisfactorily explains the different pigmentation observed from northern populations to equatorial ones, from Europe to Africa, and the lesser, but still noticeable, gradation found in Asia.

At this point we may ask ourselves what sort of pigmentation primitive man exhibited and how the present variability of pigmentation has been achieved.

The most recent discoveries seem to lend support to the interpretation that primitive humanity was dark coloured or at least mottled. In fact, it is now fairly secure that the first men originated in Africa, or at least in the equatorial regions. All primates are equatorial or subequatorial animals. In this case they certainly must have had dark skins. A light skin could have originated later as a protection against a deficiency of vitamin D. A dark skin permits only from 3% to 36% of the solar radiation to filter into the subcutaneous layer where vitamin D is synthesized from ergosterol, while a light skin will filter through from 53% to 72% of the ultraviolet rays.

When primitive man moved from the equatorial regions toward the north, he found himself living in regions where a dark skin and the need to dress himself against the cold constituted a limitation to the absorption of ultraviolet radiation. As a result there must have been frequent cases of rickets. Dark-skinned men must have been so rachitic that they would find it difficult to pursue and kill their prey, and women with darker skins must have died frequently during childbirth because of deformations of the pelvis.

Lighter-skinned individuals of both sexes would have survived better because among them rickets would have been rarer and less severe. In this way, following the classic process of natural selection, the farther north man went, the better the lighter-skinned individuals survived, while those with darker skins died or had less chance of producing offspring.

By means of a process of this sort, that is the survival of the fittest, white skin-colour was established in part of the human race.

Obviously the whole problem of skin-colour in human beings is not as simple as this, but the relation between pigmentation and hair loss during hominization is an interpretation of one of the most problematical aspects of anthropology.

GENERAL REFERENCES

Barnicot, N. A. (1957). Human pigmentation. *Man* 144, 1–7.

Barnicot, N. A. (1959). Climatic factors in the evolution of human populations. *Cold Spring Harbor Symp. quant. Biol.* 24, 115–129.

Brace, C. L. and Montagu, M. F. A. (1965). "Man's Evolution. An Introduction to Physical Anthropology", MacMillan Co./London; Collier-MacMillan, New York.

Chiarelli, B. (1963). Sensitivity to PTC (Phenyl-Thio-Carbamide) in Primates. *Folia Primatol.* 1, 103–107.

Comas, J. (1966). "Manual de Antropología física". Universidad Nacional Autónoma de México, Instituto de Investigaciones historicas, Sección de Antropología, Mexico.

Coon, C. S. (1962). "The Origin of Races", Alfred A. Knopf, New York.

Dobzhansky, T. (1962). "Mankind Evolving: the Evolution of the Human Species", Yale Univ. Press, New Haven, Conn., U.S.A.

Ehrlich, P. R. and Holm, R. W. (1963). "The Process of Evolution", McGraw-Hill Book Co., New York.

Harrison, G. A., Weiner, J. S., Tanner, J. M., and Barnicot, N. A. (1964). "Human Biology. An Introduction to Human Evolution, Variation and Growth", Clarendon Press, Oxford.

Hulse, F. S. (1963). "Human Species. An Introduction to Physical Anthropology", Random House, New York.

Ingram, V. M. (1961). "Hemoglobin and Its Abnormalities", Charles C. Thomas, Springfield, Illinois.

Kalmus, H. (1971). Phenylthiourea Testing in Primates, In "Comparative Genetics in Monkeys, Apes and Man", (B. Chiarelli, ed.), p. 65, Academic Press, London and New York.

Lasker, G. W. (ed.) (1960). "The Processes of Ongoing Human Evolution", Wayne State University Press, Detroit.

Laughlin, W. S. and Osborne, R. H. (eds) (1968). "Human Variation and Origins. An Introduction to Human Biology and Evolution", W. H. Freeman & Co., San Francisco and London.

Newman, R. W. (1970). Why Man is such a sweaty and thirsty naked animal: a speculative review. *Human Biol.* **42**, 12–27.

Parenti, R. (1954–1955). "Biologia Delle Razze Umane", Libreria Goliardica, Pisa.

4

The Concept of Species

1. INTRODUCTION

In the preceding chapters we have described the genetic character of the individual, and have seen that he is the fundamental unit of populations, while a population is the entity on which the moulding forces of nature act. These forces, in time, can lead to modifications in species by means of the diverse mechanisms of selection.

Within the limits of a population the entity superior to the individual is the "deme" which may be defined as a group of individuals organized in a self-sufficient population with its own characters. The various "demes" may display notably different physical characteristics among themselves and so represent the varieties of a species or even higher units such as subspecies.

The fundamental unit of the classification of plants and animals is, however, the species. By defining a species we arrive at the higher taxonomic units and are able to delineate the character of the lower ones. It is most important, therefore, to set forth in detail exactly what a species is.

2. DEFINITION OF SPECIES

In seeking a definition of "species" valid for the most diverse organisms, scholars in the past have used four fundamental criteria as a basis: morphological, physiological, genetic and biochemical. More recently a new criterion has been introduced: the so-called biological criterion.

The *morphological criterion* defines a species as "a group of individuals having morphological characters in common". This criterion, even though it is extremely subjective, still constitutes the most valid method of recognizing a species. However, by itself it is insufficient. The various breeds of dogs, for example, display such morphological variety that it would be difficult to attribute them to the same species if only morphological criteria were considered, while there are some species of rodents which, although morphologically very similar, certainly must be considered as diverse species.

Among the *physiological criteria*, that of cross-breeding is the clearest. By this standard every group of individuals capable of mating and producing fertile offspring is judged to be a species, while all populations whose members produce sterile or no offspring, are taken to be heterospecific. Compared with the others, this criterion has the merit of establishing a firm, constant and objectively acceptable basis of judgement; nevertheless, although obvious in itself, it has its limitations. Many good species have been found to be perfectly inter-fertile in the laboratory, while in nature, although living together in time and in space, they never or very rarely interbreed. Thus the criterion of cross-breeding, even though it is by and large the most valid, is not absolutely a definitive standard for the recognition of a species.

Among the physiological criteria there must be numbered the *ecological criterion* which has been underestimated for a long time. We shall discuss this shortly.

The *genetic criterion* is based on the possibility of obtaining among cultivated plants the isolation of subgroups whose characteristics remain constant among their descendants. It is obvious that these small species or elementary species are just assemblages of individuals having the same genotype.

One of the most recent genetic criteria for diagnosing a species is the *karyological criterion*. The number and morphology of the chromosomes within the limits of a species are in fact relatively constant. The chromosomes certainly have a notable phyletic importance, but identity of form on their part is not a sufficient criterion for ascribing two individuals to the same species.

The *biochemical criterion* endeavours to identify species from the fact that the individuals who make them up have particular biochemical characters in common. This criterion is of special importance in organisms like the Bacteria, which lack, entirely or almost so, morphological criteria, but it is difficult to apply it in the case of the Mammals where similar organic substances may have originated through phenomena of adaptive convergence or may have been diversified by single mutations.

The so-called *biological criterion*, of which Mayr is the principal author, includes and supersedes the preceding but also takes into consideration besides the others the aspect of "behaviour", which in all vital manifestations unites the individuals of one species and distinguishes them from others.

This criterion, which contemporaneously makes use of other criteria for diagnosing species, would be better named "synthetic" or "multilateral" than biological. It is the criterion which is most used in practice and has led to a definition of what is meant by a species consonant with reality and with the experimental and theoretical requirements of today.

The biological criterion assumes that each distinct species (*bona specie*) of animals or plants which reproduces sexually does not form hybrids with other species cohabiting in the same area, and that within its ecological niche it has requirements so different from those of any other that they can co-exist in the same locality. Therefore, according to Mayr, "species are groups of inter-fertile populations reproductively isolated from other similar groups".

3. Biological Concept of Species

Thus this definition of species postulates a cohesion among the individuals who form a single species determined by three factors:

(1) the ability of the members of a species to discriminate and recognize each other as belonging to a single reproductive community;

(2) the genetic cohesion among the components of a single species, determined by the continuous mixing of the chromosomal material, so that the species (or more precisely the reality of the species) is none other than a discontinuous complex of the genes of the populations of which the species is composed;

(3) the ecological interaction of this species with other species of plants and animals.

This concept of species is based on the presupposition that species are composed of populations and that the characteristics of a species are not typological but statistical. Therefore, because of this statistical aspect of the characters of a species, statistical averages can be substituted for a typological description. Thus it is necessary to know whether the populations being examined produce fertile offspring if they are cross-mated, and to consider the complex factors which contribute to the isolation between species.

In practice, therefore, in order to decide whether two populations or groups of individuals represent two different species or not, it is necessary to dispose of three types of information:

(1) whether from the morphological point of view the populations are identical or not;

(2) whether they are reproductively isolated or not;

(3) whether they co-exist in the same area and are therefore sympatric, or do not co-exist in the same area and are, therefore, allopatric.

These three kinds of information give rise to the following eight possibilities:

(1) if the two forms are morphologically identical, are sympatric and not isolated reproductively, they naturally belong to the same population;

(2) if the two forms are morphologically identical and allopatric and they are reproductively isolated, they belong to the same population;

(3) if the two forms are morphologically different but sympatric and not isolated reproductively, they are genetic variations within the same population: that is they represent a case of polymorphism;

(4) if the forms are morphologically different and allopatric but not reproductively isolated, they belong to subspecies of a single species;

(5) if the two forms are morphologically identical and sympatric but reproductively isolated, they represent two *bonae species* but are so similar as to justify the use of a particular expression, that of "twin species";

(6) if two morphologically identical but allopatric forms are also reproductively isolated, they represent two different, even if very similar, species (twin species);

(7) if the two forms which are reproductively isolated are also different in morphology and live in the same area, they belong to different species;

(8) if the forms are morphologically different, allopatric and reproductively isolated they belong without doubt to different species.

Such an outline (Table 4.1), even if oversimplified, may be useful in

TABLE 4.1. A possible taxonomic breakdown. For cases 6 and 8 the existence of intermediary populations is assumed.

The individuals are:		Reproductively non-isolated	Reproductively isolated
Morphologically identical	Sympatric	1. The same population	5. Twin species
	Allopatric	2. The same population	6. Twin species
Morphologically different	Sympatric	3. Individual variation of the same population (polymorphism)	7. Different species
	Allopatric	4. Different subspecies	8. Different species

defining all the possible situations with which a taxonomist might have to deal.

Biologically, species can be defined as a system which assures and protects favourable genetic combinations. In fact the possibility of mating, and thus an exchange of genes, ensures both for the individual and the population the variability necessary for adaptation to diverse conditions. On the other hand, the impossibility of external (extra-specific) mating protects the group from the introduction of unsuitable characters which would upset or at least disturb the dynamic equilibrium of the genes already established.

4. Subspecies, Varieties and Population Groups (Races)

The taxonomic division immediately below species is the subspecies. By subspecies we mean a local population of a polytypical species which is distinguished from the other populations of the same species by occupying a distinct geographical territory and by considerable qualitative or quantitative differences.

Subspecies must therefore be, by definition, allopatric; in fact, if several subspecies should live in the same region they would mix genetically and the differences between them would disappear. The differences from the other subspecies of the same species are kept constant by geographical isolation, and the subspecies must be specially adapted to the climate and the physical conditions of the environment in which it is found by means of physiological characteristics.

Within the limits of the populations which constitute a subspecies, complexes of genes by which certain populations differ from others of the same subspecies may be established through the action of mechanisms of various kinds (genetic drift, geographical or social isolation and so on). To these population groups the name of "race" is generally given. This unit must, therefore, be considered as the taxonomic unit below that of subspecies. As such, it can be distinguished only in a polymorphic species such as the human species, that of the dog, and of many other domestic animals. Its definition is empirical: it is based on morphological, anatomical, physiological, genetic and psychological characters and might be defined as: "a group of individuals in the population of a species living in a definite area or areas, which, because of an assemblage of physical characteristics common to its members but variable within certain limits, can be distinguished from other groups within the species".

General References

Blackwelder, R. E. (1967). "Taxonomy. A Text and Reference Book", J. Wiley and Sons, New York, London and Sydney.

Mayr, E. (1963). "Animal Species and Evolution", The Belknap Press of Harvard University Press, Cambridge, Mass., U.S.A.

Washburn, S. L. (ed.) (1963). "Classification and Human Evolution", Methuen & Co./Wenner-Gren Foundation for Anthropological Research, London.

5

The Living Primates

1. GENERAL

The Order of Primates, to which man belongs, was defined in 1758 by
Linnaeus and included: man, monkeys, lemurs, bats and the tardigrades.
Later these last two groups were classified in other orders.

Today the Order of Primates includes: the prosimians, (Tupaiidae,
Lorises, Tarsiers and Lemurs), the American or platyrrhine monkeys, the
Old World or catarrhine monkeys, the anthropoid apes and man.

The name of "Primates" was given by Linnaeus to the monkeys and man,
to distinguish them from the other mammals which he called "Secundates",
and from all the other animals to which he gave the name "Tertiates".

The features which characterize the primates and distinguish them from
the other mammals are:

(1) The primates are for the greater part arboreal, with the exception of
the baboons, the macaques, the patas, the gelada and man, who live
almost exclusively on the ground. The arboreal way of life is evidently
an archaic one: the primitive placentals were arboreal and the primates
continue to be so. As a mode of existence it is accompanied by precise
anatomical and physiological modifications.

(2) They climb by grasping the branches with prehensile limbs, having exceptionally flexible arms and legs.

(3) The mammals in general and all the primates, except man, habitually move on all four feet; in the primates, however, life in the trees has produced a differentiation between the anterior and posterior limbs. The former are predominantly used for climbing and the latter for support.

(4) The long primitive tail has been preserved by some primates as an aid in maintaining equilibrium, and by others with a prehensile function; in other forms the external tail has been reduced or is completely missing.

(5) Because of their characteristic way of life, the primates have preserved the generalized structure of the limbs typical of primitive mammals. In fact all, or almost all, the parts of the "acropodium" are present, and elements of the thoracic skeleton, such as the clavicle, which in many other groups of mammals tend to disappear or to be much reduced, have been preserved and well differentiated.

(6) The primates have exceptionally mobile limbs and have acquired an independent mobility of the digits, especially of the index finger and of the thumb, which can be opposed to the other fingers in some primates.

(7) In the primates the sharp claws typical of the mammals have been substituted by flat nails and at the same time the soft part of the last phalanx of the finger has developed greater sensitivity.

(8) The primates are for the greater part omnivores. Their dentition is, therefore, little specialized: they lack the great development and specialization of the canines typical of the carnivores or that of the molars typical of the herbivores. Moreover, the number of teeth has been reduced from 44 (as it was in the primitive placentals) to 32 (or 36), yet the conformation of simple cuspids has been preserved in the molars.

(9) In relation to their body-weight the primates have the most highly developed brain of all the mammals. In all Primates, in fact, the brain–body-weight ratio is not allometric though it is for all the other organs. The general increase in size is associated with a greater complexity of structure. The dimensional increase affects the motor area particularly and, therefore, is probably related to the higher degree of muscular co-ordination necessary for an arboreal life, to the complex movements of the upper limb and to the varying degrees of tendency to binocular vision.

(10) As a consequence of the reduction in the number of teeth, of the lack of specialization in the masticatory apparatus, of the improvement of vision, of the progressive reduction of the olfactory apparatus and of the increase in the size of the brain, the structure of the primate cranium definitely differs from that of the other mammals. The face has a

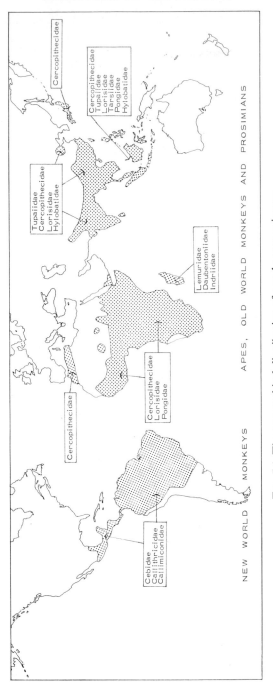

FIG. 5.1 The geographical distribution of non-human primates

proportionally smaller development, and above all is shorter; the mouth is much retracted and the profile is more perpendicular.

Moreover, the orbits do not communicate with the temporal fossa but constitute a clearly defined concave cavity (retro-orbital closure), which in the higher primates approaches a hemisphere. This is a typical characteristic of the primates, which is found in the most ancient forms and which differentiates this group from the other types of mammals.

(11) The primates are, moreover, characterized by a long period of pre-natal life and by a most efficient placental-foetal rapport which improves phylogenetically from the most primitive to the most differentiated.

(12) Another general trend in primates is a reduction of the number of offspring, aiding their adaptation to the arboreal life. The arboreal habitat forces the reduction of the number of embryos developing in the mother's body and, after birth the number of babies that can be carried. This reduction helps the development of a stronger mother–infant bonding which is an important prerequisite for hominoid evolution (learning process and socialization).

Not only does this series of characters taken as a whole serve to differentiate the living primates from the other mammals, but they are also valid for the study of primate fossil remains in that most of them involve particular characteristics or transformations of the skeleton.

2. THE CLASSIFICATION OF THE PRIMATES

Within its limits the Order of Primates presents a remarkable variety of forms, which may be classified into approximately two hundred species and subdivided into about fifty genera. In order to understand the evolution of this group of animals more clearly and to classify our species correctly within the Order it is essential to organize the various forms into groups having a taxonomic value intermediate between orders and species: sub-orders, infra-orders, superfamilies and genera.

There is still some disagreement among taxonomists regarding the classification of the various forms of primates. Most of these difficulties concern the categories superior to species and genera.

Some of these difficulties arise from the direct involvement of man in the system that he studies; Simpson has made an apt remark about this: "The peculiar fascination of the primates and their publicity value have almost taken the order out of the hands of sober and conservative mammalogists and have kept, and do keep, its taxonomy in a turmoil" (Simpson, 1945).

The classification which is presented here (Table 5.1) is drawn principally from Simpson, with some modifications. With the exception of man the various living species of primates mainly inhabit equatorial regions (Fig. 5.1).

3. LIVING PROSIMIANS

In this suborder the systematists include the lemurs, the lorises, the tarsiers and, although not everyone is in agreement, the tupaia. As a group the prosimians are particularly interesting in that they are for the most part very primitive mammals and have many characters in common with the mammalian group closest to the primates: the insectivores.

The ancestors of the prosimians constituted the fundamental branch from which all present-day primates have evolved. The knowledge which we have of them helps us to understand the relationships found among the various forms of existing primates.

TUPAIOIDEA. These form an aberrant group of prosimians which many authors prefer to classify with the insectivores. The group consists of very primitive forms, represented by the genera *Tupaia, Dendrogale, Urogale, Ptilocercus*. All are confined to the Far East, more precisely to southeast Asia and Indonesia (Fig. 5.2). They are among the smallest existing mammals, very lively and aggressive, and as their metabolic rate is very high they eat voraciously. They are omnivorous, but eat insects very willingly. Like the insectivores they have claws on all ten digits.

FIG. 5.2 *Tupaia glis* (courtesy of Tierbilder Okapia).

They have a total of 38 teeth, distributed according to the formula: $\dfrac{2I, 1C, 3P, 3M}{3I, 1C, 3P, 3M}$. These animals are the only primates which have six lower incisors; all the others have two or four. Their eyes are large and are placed at right angles to each other so that their vision is not stereoscopic as in the other primates. Nevertheless, the visual area of the brain (area striata) is larger than that of the insectivores and the olfactory area is relatively smaller.

The internal skeleton is lemuroid in type, particularly as regards the rounded orbit and the swollen *bulla tympani*.

The brain in general is primitive although there is a slight development of the "neopallium". The male has a large, pendulous penis and external testicles, the female has a bipartite uterus and the embryo is implanted with a discoidal, endothelial, chorial placenta having two discs. The placenta of *Tupaia* is therefore unique among Mammals in showing certain structural advances towards the hemochorial type characteristic of all the Primates. The genus *Tupaia* is characterized by having three pairs of breasts, the genus *Dendrogale* two and the genera *Urogale* and *Ptilocercus* one pair only.

Remains of fossil *Ptilocercus*, not very different from the present form have been found in the lower Oligocene layers in Mongolia.

LEMUROIDEA. The superfamily of the Lemuroidea comprises three families (Lemuridae, Indriidae and Daubentoniidae) actually living exclusively in Madagascar. On this island, which has been separated from the continent over a long period and where no other type of mammal has ever competed with them, the lemurs have specialized like the marsupials in Australia. As has happened to many other examples of isolated fauna, many larger species of lemur disappeared by the arrival of man about 1000 B.C. when the island was opened up to continental animals.

Most of the lemurs have a skeleton especially adapted to climbing trees, and they display claws on the second or third digits of their hands and feet, while the other digits have nails like those of the other primates. In some species the second digit is rudimentary, and in almost all species the fourth is the longest, rather than the third as is found in the other primates. The mandible is long and slender, V-shaped, and compared to the Tupaioidea, contains one tooth less on each side. The dental formula is: $\dfrac{2I, 1C, 3P, 3M}{2I, 1C, 3P, 3M}$ with a total of 36 teeth, organized, however, in a peculiar way. In the mandible the four incisors and the two canines are sharp and converge externally, the first premolar is large and assumes the functions of a canine.

An even greater degree of specialization is seen in *Daubentonia*, whose dentition is extremely reduced; the dental formula is: $\dfrac{1I, 0C, 1P, 3M}{1I, 0C, 1P, 3M}$'

TABLE 5.1. Taxonomic classification of the living primates according to recent opinion.

SUBORDER	INFRAORDER	SUPERFAMILY	FAMILY	SUBFAMILY	GENERA	NO. OF SPECIES
PROSIMII	TUPAIIFORMES	TUPAIOIDEA	TUPAIIDAE	TUPAIINAE	*Tupaia*	9
					Dendrogale	2
					Urogale	1
				PTILOCERCINAE	*Ptilocercus*	1
	LORISIFORMES	LORISOIDEA	LORISIDAE		*Loris*	1
					Nycticebus	2
					Arctocebus	1
					Perodicticus	1
			GALAGIDAE		*Galago*	6
	LEMURIFORMES	LEMUROIDEA	LEMURIDAE	LEMURINAE	*Lemur*	5
					Hapalemur	2
					Lepilemur	1
				CHEIROGALEINAE	*Cheirogaleus*	2
					Microcebus	2
			INDRIIDAE		*Indri*	1
					Avahi	1
					Propithecus	2
			DAUBENTONIIDAE		*Daubentonia*	1
	TARSIIFORMES	TARSIOIDEA	TARSIIDAE		*Tarsius*	3
			CALLITHRICIDAE	CALLITHRICINAE	*Callithrix*	3
					Leontideus	3
					Cebuella	1
				CALLIMICONINAE	*Callimico*	1

ANTHROPOIDEA

- PLATYRRHINI
  - CEBIDAE
    - AOTINAE
      - Aotus — 1
      - Brachyteles — 1
      - Callicebus — 3
    - PITHECINAE
      - Pithecia — 2
      - Chiropotes — 2
      - Cacajao — 3
    - ALOUATTINAE
      - Alouatta — 5
    - CEBINAE
      - Saimiri — 2
      - Cebus — 4
    - ATELINAE
      - Ateles — 4
      - Lagothrix — 2
- CATARRHINI
  - CERCOPITHECOIDEA
    - CERCOPITHECIDAE
      - PAPIINAE
        - Macaca — 13
        - Cercocebus — 5
        - Papio — 7
        - Theropithecus — 1
      - CERCOPITHECINAE
        - Cercopithecus — 22
        - Erythrocebus — 1
    - COLOBIDAE
      - COLOBINAE
        - Presbytis — 14
        - Pygathrix — 1
        - Rhinopithecus — 2
        - Simias — 1
        - Nasalis — 1
        - Colobus — 5
  - HOMINOIDEA
    - HYLOBATIDAE
      - Hylobates — 6
      - Symphalangus — 1
    - PONGIDAE
      - Pongo — 1
      - Pan — 2
      - Gorilla — 1
    - HOMINIDAE
      - Homo — 1

FIG. 5.3 *Nicticebus coucang* (courtesy of B. Grzimek).

that is with a total of 18 permanent teeth, compared with the 36 of the other lemurs. Moreover, the incisors of this species are sharp and grow exactly as do those of rodents. In fact, they live on juice sucked from the hard stems of sugar cane and such highly differentiated incisors are, therefore, essential.

Unlike most of the other primates, the lemurs have a seasonal reproduction. The uterus of the lemur is bipartite like that of the Tupaioidea, but where the placenta of the latter is disc-shaped, that of the lemur is diffuse with a wide allantois.

Another feature unique among the primates is that the placenta is not

FIG. 5.4 *Galago senegalensis* (courtesy of B. M. Solandt, Toronto).

expelled at the moment of birth. In this characteristic the lemurs very much resemble the ungulates.

Almost all the genera display a single pair of breasts located in the pectoral region.

In their dimensions they vary from the size of a rat to almost that of a man if we include the sub-fossil form. They have a fairly well developed sense of smell and a comparatively small brain.

Although they are specialized in a remarkable number of forms, their intellectual capacity has not increased during their evolution from the time they arrived in Madagascar. This fact must certainly be considered in relation to the lack of competitors on the enormous island.

The most typical and representative living members of this superfamily are the various species of *Lemur*, of *Microcebus*, of *Indri*, and of *Daubentonia*.

Altogether they are a highly specialized group and are, therefore, not very useful for furnishing information on the origin of the primates; still they are very important for illustrating some aspects of specialization and some paths of evolution.

As we shall see, fossil remains which may be ascribed to these forms have been found in the sediments of the Palaeocene and Eocene in Europe, North

America and Africa. They have been assigned to the families Plesiadapidae, Adapidae and Notharchidae.

LORISOIDEA. The superfamily Lorisoidea comprises two families: that of the Lorisidae with four genera and that of the Galagidae with only one. The Lorisidae include species living in southeast Asia and Africa. They are all small animals, the size of a squirrel or at the most of a cat, with thick fur, a round head and short ears; the tail rudimentary or absent; nails on all digits except the second. They are nocturnal and solitary. They move slowly. Their diet in general consists of leaves and birds' eggs. The "bush baby" (*Galago*) is by contrast more mobile; it lives in the trees and moves about by agile leaps. (Fig. 5·4).

TARSIOIDEA. The superfamily Tarsioidea is represented by only one living genus and probably by only one species: *Tarsius spectrum*. It is, however, extremely important as it is thought that the fossil forms found in Palaeocene and Lower Eocene strata in France and North America represent a link between the lemurs and the other more evolved primates. All these fossil forms have been placed in the family Anaptomorphidae.

The superfamily, Tarsioidea, as we have said, includes the only living form *Tarsius*, which lives in the forests of Sumatra, on the Celebes and Sangihe groups of islands and in the southern part of the Philippines. The various populations of this species vary in colour and in some small characteristics of the cranium. They are very small animals, and are equipped with enormous eyes that are frontally disposed and therefore give a version almost, but not perfectly, stereoscopic. Both their hands and feet are covered, on their inner aspects, by wide spongy cushions. They are further characterized by an elongation of the tarsal bone, from which their name is derived. The dental formula is: $\dfrac{2I, 1C, 3P, 3M}{1I, 1C, 3P, 3M}$ with a total of 34 teeth. In the jaw the median incisors are of the same size as the lateral ones, as in man.

While in the lemurs and the Lorisidae the osseous part of the ear is composed of two parts (the *bulla* and *anulus*), in *Tarsius* these are fused to form a single external meatus, as in the higher primates.

4. GENERAL CHARACTERISTICS OF THE PLATYRRHINE
AND CATARRHINE MONKEYS

Characteristics common to all the monkeys, the anthropoid apes and man but different from the prosimians are: the orbital cavity separated from the temporal fossa by means of an osseous septum; the *bulla tympani* not swollen; a highly developed brain with numerous convolutions; the uterus single and rounded; the placenta formed of embryonic and maternal tissues, disc-shaped and expelled at birth (decidual).

FIG. 5.5 *Callithrix jacchus*
(courtesy of Hannover Zoo).

(b)

FIG. 5.6 (a) *Pithecia pithecia* ♂ and (b) ♀ with
young (courtesy of Cologne Zoo).

(a)

Fig. 5.7 *Callimico goeldii* (black) with a Callithricinae (courtesy of Rühmekorf, Hannover).

5. Living Platyrrhine Monkeys

The name of this group of primates, which live in Central and South America, is derived from the fact that the nostrils are separated by a wide cartilaginous septum and open laterally, in contrast to the Old World monkeys, or catarrhines, whose nostrils are very close together and open down. The head is rounded, with a relatively small mandible. A further difference from the catarrhines is that they do not have ischial callosities. The thumb is never opposable to the other digits and is often vestigial or absent. The big toe is equipped with a flattened nail. The nails of the other digits, of both hands and feet, are often compressed laterally and in the

Callithricidae they appear still to be proper claws, although on histological examination they have more the characteristics of nails.

The American monkeys are all arboreal and live in the tropical forests of South America. Only a few species extend into Central America.

Their general dental formula is lemuroid; that is, composed of 36 teeth disposed thus: $\dfrac{2I, 1C, 3P, 3M}{2I, 1C, 3P, 3M}$. They are divided into two families: Callithricidae and Cebidae.

The most primitive forms are those of the Callithricidae with the genera *Callithrix*, *Cebuella* and *Leontideus*. They are of extremely small dimensions, with a tail as long as the head and body together (Fig. 5.5). The dental formula of the Callithricidae is different from that of the Cebidae and is reduced, with two molars instead of three in each side of the jaw.

The Cebidae represent a more evolved form, having nails on all digits instead of just on the big toe (Fig. 5.6a, b). Vision is stereoscopic, as in the catarrhine monkeys. They have a well-developed brain and an intelligence superior to that of the Callithricidae. The locomotion of the Cebidae is also highly differentiated; a peculiar characteristic of some genera is the long, prehensile tail used as a fifth limb, a specialization that is particularly advantageous during the rainy season when the forests are flooded.

Between the Cebidae and the Callithricidae there is an intermediate form represented by the species *Callimico goeldii* (Fig. 5.7); indeed *Callimico*, *Callicebus* and *Cebus* form a progressive succession of forms which resembles a fossil evolutionary series much more than a systematic succession of living forms. In fact, it is considered that *Callimico goeldii* has still retained many ancestral characteristics of the group which, divided from the remaining species of primates at the end of the Tertiary, gave rise to the various forms of primates at present living on the American continent.

6. Catarrhine Monkeys

All the monkeys of the Old World (Asia, Africa and Europe), the anthropoid apes and man are classified together under the name Catarrhine. They have in common the type of nose from which they take their name: in it the nostrils are narrow and parallel, separated by a narrow nasal septum. Their sense of smell is relatively low compared with the other primates. They are all diurnal and have stereoscopic vision.

They have a total of 32 teeth set out according to the formula: $\dfrac{2I, 1C, 2P, 3M}{2I, 1C, 2P, 3M}$. Within the catarrhines group, it is possible to differentiate groups having common characteristics. One such characteristic, for example, is the type of placenta, which in the gibbons, apes and man consists of a single

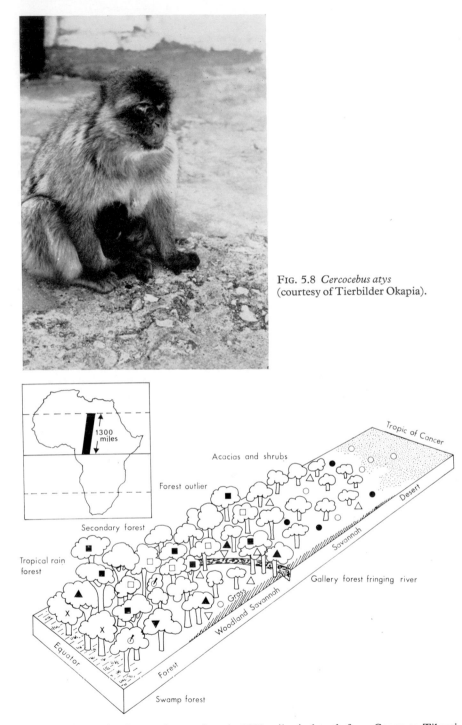

FIG. 5.8 *Cercocebus atys*
(courtesy of Tierbilder Okapia).

FIG. 5.9 Vegetational zones in sample strip 1300 miles in length from Congo to Tibesti plateau showing resident primates. O *Papio anubis*; ● *Erythrocebus patas*; △ *Cercopithecus aethiops*; ▲ *Cercopithecus mona*; □ *Cercopithecus nictitans*; ■ *Colobus badius*; ▽ *Cercocebus atys*; ▼ *Cercopithecus neglectus*; × *Cercopithecus nigroviridis*; ♂ *Cercopithecus cephus*.

FIG. 5.10 *Macaca sylvana*
(courtesy of F. Burton, Toronto).

FIG. 5.11 *Cercopithecus nictitans*
(courtesy of B. Grzimek).

disc, but in the other catarrhines is formed of two discs. Another characteristic which differentiates the apes and man from the other catarrhines is the absence of ischial callosities (moist surfaces lacking hair which cover the protrusion of the ischial bones). These callosities are still sometimes present in the chimpanzee.

Although, as we shall see, many problems concerning the taxonomic organization of the group exist, the catarrhines can be subdivided into the superfamilies Cercopithecoidea and Hominoidea. In the first superfamily we may include, even if the majority of the systematists are not in agreement, the families of the Cercopithecidae, the Colobidae and the Hylobatidae; in the second superfamily the true apes (Pongidae) and man (Hominidae).

In the following pages we shall describe the principal characteristics of the different families.

7. THE LIVING CERCOPITHECOIDEA

The family of the Cercopithecidae includes the macaques, the baboons and the cercopithecines that of the Colobidae includes the Colobus monkeys, *Presbytis*, *Nasalis* and other forms.

All the representatives of the Cercopithecidae are characterized by buccal sacs in which they can store food temporarily. Their stomach is simple: they are omnivorous. Their geographical distribution is extensive: the macaques are distributed principally over all of India, in the north of China and in Japan (Fig. 5.8); only one of this species, the Barbary ape, lives in Africa and in Europe (Gibraltar). The other Cercopithecidae live in Africa, either in the forests, the savannah or in the mountains of Ethiopia (Fig. 5.9). Among these, however, the genera *Papio*, *Theropithecus* and *Cercocebus* must be clearly distinguished from the species of the genus *Cercopithecus*; in this last genus, in fact, among other differences the last lower molar has only four cusps instead of five. This difference is considered to be very ancient.

The locomotion of the Cercopithecidae depends on the habitat which distinguishes the two groups, whether arboreal or terrestrial. The *Cercocebus* (Fig. 5.10) and *Cercopithecus* genera (Fig. 5.11), which live in Africa, are prevalently arboreal.

The other genera (*Macaca*, *Papio*, *Theropithecus*) are instead specialized for a way of life unknown in the primates thus far described: life on the ground. By this means they have been able to colonize environments new to animals such as the primates; that is, rocky areas or those of the savannah or the plain. They are able to sit erect during moments of rest and have retained the complete mobility of their digits in spite of their quadrupedal

Fig. 5.12 *Presbytis obscurus*
(courtesy of B. Grzimek).

Fig. 5.13 *Colobus guerezza*
(courtesy of B. Grzimek).

comportment. They have strong jaws and large teeth. They are able to open their mouths completely only when their heads are tilted back.

They live in groups under the leadership of an old and strong male, or, in his absence, by the oldest female.

The females of many Cercopithecidae have a large area around the genital region which becomes swollen and vivid during the period of ovulation and so serves as a sexual attraction. The male often has his sexual organs gaudily coloured in green, red and blue.

The family of the Colobidae includes species which live in Africa and southern Asia, all of which are arboreal.

The genus *Colobus* lives in Africa (from West Africa along the Congo and into Ethiopia); on the other hand other genera (*Presbytis, Nasalis, Simias, Pygathrix, Rhinopithecus*) live in Asia (Figs 5.12, 5.13 and 5.14). Their diet is based on leaves, buds, small shrubs and fruit. This alimentary adaptation has made a special digestive apparatus necessary, and the stomachs of the Colobinae are extremely large so that they can hold a quantity of leaves equal to one-third of the animal's weight; they are also subdivided into several semi-independent sacs. For this reason the Colobidae do not have the buccal sacs which are present in the Cercopithecidae. Because leaves require a particular type of mastication, the incisors of these animals are small and sharp. The molars are proportionally much larger. The grinding mechanism is a product of side-to-side movements, which leads to a remarkable development of the masseters and the temporal muscles of the cranium.

Because of this special buccal apparatus and complex musculature, many of these species bear resemblances to the Hominidae. Some of them (*Nasalis* and *Rhinopithecus*) with particularly well developed external noses can be considered complete caricatures of human beings.

The exclusively vegetarian diet of the Colobidae which affords them abundant and easily found food, nevertheless limits them to a habitat in the region of forests with non-deciduous leaves. Some of these species are also able to live in cold areas, such as the Himalayan and trans-Himalayan forests (*Presbytis* and *Pygathrix*). However, as for other ecologically limited specializations, their diet, from an evolutionary point of view, constitutes a blind alley which more than one group of animals has entered but from which *none* has emerged to produce more evolved forms.

By reason of some of their characters, such as the ability to stand erect and the absence of a tail, the Hylobatidae have in the past been classified in the superfamily Hominoidea. More recent knowledge tends rather to place this group among the Cercopithecoidea. As we have said the Hylobatidae (gibbons) are characterized by the absence of a tail and by very long limbs. Among the primates they are the best brachiators and are able to leap with

FIG. 5.14 *Nasalis larvatus*
(courtesy of Tierbilder Okapia).

FIG. 5.15 *Hylobates concolor*
(courtesy of Zoologischer garten
Hannover).

FIG. 5.16 *Pongo pygmaeus*
(courtesy of Tierbilder Okapia).

FIG. 5.17 *Gorilla gorilla*
(courtesy of B. Grzimek).

great agility from branch to branch and from one tree to another using only their upper limbs.

They also succeed in walking erect, balancing themselves with their long arms; their hands and feet are long and slender and are used indifferently for seizing food. The thumb is short and attached far back on the palm of the hand so that it is almost completely unusable. This, as we shall see, is a particular adaptation to brachiation.

Their teeth are small, except for the canines which are large and long. They utter shrill sounds which can be heard a kilometre away. They do not display a visible sexual dimorphism as the genitals are small and covered by hair.

The Hylobatidae comprise two genera: *Symphalangus* and *Hylobates*.

Symphalangus is characterized by a membrane between the first phalanges of the second and third digits, from which the name is derived, and a laryngeal sac which is extensible in a particular way. It inhabits the forests of Sumatra.

Hylobates on the other hand, lives especially on the Malay Peninsula around Malacca, in Borneo and Sumatra (Fig. 5.15). In the past this genus and similar genera had a much more extensive geographical distribution, as the remains of gibbons found in Miocene strata in Europe and in the Pliocene of central and northern China attest.

8. Living Anthropoid Apes

The family Pongidae includes the three true apes: Orangutan, Gorilla and Chimpanzee (Figs. 5.16, 5.17 and 5.18).

The orangutan lives on the islands of Borneo and Sumatra, while the other two apes live in the forests of Africa. Anatomically and physiologically the chimpanzee and the gorilla are very similar, being differentiated more by subtle variations than by concrete characteristics.

The differences between the apes and man are largely correlated with habitat. Man has become terrestrial whereas the apes have kept their arboreal habits. These habits have sometimes resulted in external developments, as in the orang which has extremely long upper limbs and relatively short lower ones. The gorilla, especially the male, has become partly terrestrial because of its great size. The chimpanzee is indifferently arboreal or terrestrial.

The disproportion between the long upper limbs and the short lower ones has produced in the gorilla and chimpanzee a most peculiar way of moving on all fours, with the anterior part of the body more elevated and the weight resting on the knuckles of the hands. The shape of the pelvis and

FIG. 5.18 *Pan troglodytes* (courtesy of Zoological Garden, Turin).

the small size of the gluteal muscles deny to the apes an habitually erect posture.

An examination of the menstrual cycle of apes reveals that the human female and the chimpanzee have the same phase and that the intervals of the cycle are also the same. The gestation period of the three apes is of about nine months and the placenta is of essentially the same type as the human placenta.

The brain too, although it is only about one-third of the size of the human brain, is virtually a miniature of it.

There is not a part, not an organ of man that is missing in the apes; the differences consist only in diverse proportions. Another indication of the similarity between man and the apes is given by the susceptibility of the latter to diseases specific to man, while the other primates are relatively immune.

9. THE HOMINIDS

The family Hominidae is represented by a single living genus and only one species, *Homo sapiens*.

Our species differs from the other apes by easily enumerated characters. Man's brain is about three times as large as that of the apes, although at birth it is smaller than that of the orangutan. It is also more specialized. The teeth of human beings are generally smaller and, in particular, the canines are of the same size as the incisors and premolars. Consequently the diastema, that is a space between the canines and the lateral incisors which is present in all the apes, is not found in man except in rare instances.

While the apes are specialized as brachiators, man has acquired an erect posture with consequent important differences in the spinal column, the pelvis, the upper and lower limbs, the hands, the cranium, the mandible and in the special development of the buccal apparatus. The majority of human populations are less hairy than the apes except for the head, axilla and pubic region.

10. SPECIFIC UNITY BETWEEN LIVING HUMAN BEINGS (*Neanthropus*) AND THE FIVE HUMAN SUBSPECIES

Living man is distributed over practically all the surface of the earth; that is, he is cosmopolitan. On external morphological examination he is found to display great variety: belonging to the species are groups of blond individuals with blue eyes, very fair skins, and tall stature (average 182cm) like the Swedes, and groups of individuals having brown skins, very short, black, tightly curled hair, and extremely short stature (average 144cm) like the African Pygmy. Notwithstanding these external morphological differences, all the various human groups belong to one species, *Homo sapiens*.

In fact, among existing peoples practically every possible cross-mating has been recorded. It has almost always been the Europeans who, spreading to all parts of the world and coming in contact with other populations, have served as the test of cross-fertility. There is evidence of mating between Europeans and Tasmanians (the *Salers* of the islands in Bass Strait), between Europeans and Hottentots (the bastards of Rehoboth), between Negroes and Europeans, between Hottentots and Indians in South Africa, between Ainu and Japanese, between Scandinavians and Chinese, and no instance of collective sterility has ever been recorded. Therefore, according to the condition of interfertility which we have considered implicit in the definition of a species, all human populations must be grouped into only the categories below that taxonomic unit. At least this holds if the validity of the derivative premise is admitted; that is if the interfertility of two groups is

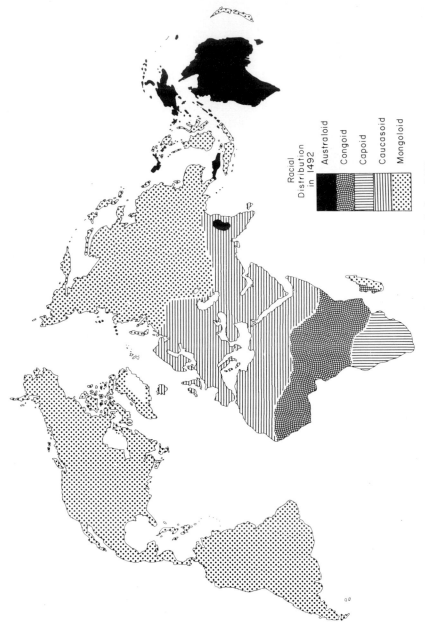

Fig. 5.19 Map of the distribution of the five human subspecies before 1492.

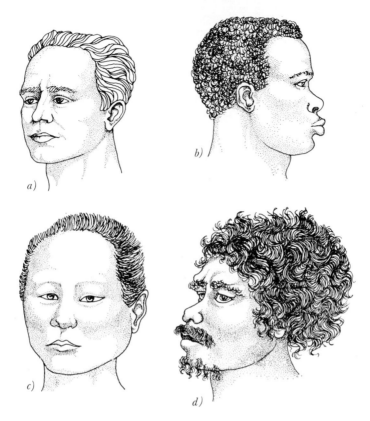

FIG. 5.20 Some variations in human facial characteristics: (a) Caucasoid; (b) Congoid; (c) Mongoloid; (d) Australoid (from Bates 1961).

considered to be demonstrated by their mutual interfertility with a third group. Thus, if we turn back to the outline of possible taxonomic conditions on page 51, we find that individuals belonging to human populations fall into groups which are not reproductively isolated, even if they are morphologically different or allopatric, and, therefore, they all belong to the same species (headings 1, 2, 3 and 4).

However, within the limits of the populations which make up the species "*Homo sapiens*" there are some that, even though they are interfertile with the others, are morphologically different and up to the fifteenth century lived in different regions: Negroes and Whites, Australian aborigines and Negroes, and so on.

On the basis of these criteria it is possible, according to Coon, to distinguish five diverse subspecies among human beings (Figs 5.19 and 5.20):

(1) Caucasoids: distributed geographically in all of Europe, in Arabia, the Middle East, India and North Africa.

(2) Mongoloids: distributed geographically in all of Asia, the Philippine Islands, North, Central and South America, and the eastern part of Madagascar.

(3) Australoids: distributed geographically in all of Australia, Borneo, the central part of India and some coastal regions of the Philippines.

(4) Congoloids: distributed geographically in central Africa and in the western part of Madagascar (Malagasy Republic).

(5) Capoids: distributed geographically in southern Africa.

Further, it is found that populations which co-exist in the same region may be rather diverse morphologically, and, therefore, within the compass of the five human subspecies described above, groups of diverse populations may be distinguished. The name of "race" is generally attributed to such groups.

General References

UNESCO Committee work (1956). "The Race Question in Modern Science", UNESCO, Paris.

Bates, M. (1961). "Man in Nature", Prentice-Hall, Englewood Cliffs, New Jersey.

Buettner-Janusch, J. (1966). "Origins of Man. Physical Anthropology", J. Wiley & Sons, New York, London and Sydney.

Chiarelli, B. (1972). "Taxonomic Atlas of the Living Primates", Academic Press, London and New York.

Cole, S. (1965). "Races of Man", British Museum (Natural History), London.

Coon, C. S. (1962). "The Origin of Races", Alfred A. Knopf, New York.

Eimerl, S., De Vore, J. and the Editors of *Life* (1966). "The Primates" (*Life* Nature Library), (Netherlands), Time-Life Int.

Garn, S. M. (1961). "Human Races", Charles C. Thomas, Springfield, Illinois.

Hill, W. C. O. (1953). "Primates: Comparative Anatomy and Taxonomy", Vols 1–8, Edinburgh University Press, Edinburgh.

Hofer, H., Schultz, A. H., and Starck, D. (Hg.) (1956). "Primatologia Handbuch der Primatenkunde", Vol. 11, Systematik Phylogenie Ontogenie, S. Karger, Basel and New York.

Kephart, C. I. (1960). "Races of Mankind", Philosophical Library, New York.

Martin, R. D. (1968). Towards a new definition of primates. *Man* **3**, 377–401.

Napier, J. and Barnicot, N. A. (eds) (1963). "The Primates" (Symposia of the Zoological Society of London, Vol. 10), Zoological Society of London, London.

Napier, J. R. and Napier, P. H. (1967). "A Handbook of Living Primates: Morphology, Ecology and Behaviour of Non-human Primates", Academic Press, London and New York.

Schultz, A. H. (1969). "The Life of Primates", Weidenfeld and Nicholson, London.

6

Evolution of Species (Speciation)

1. GENERAL

In the preceding chapter we have briefly reviewed the object of our attention, the various species of primates at present living, and have defined them as relatively stable entities. However, their dynamic aspect is of greater interest; that is, how these various species of primates have become differentiated, and how they could have evolved from a common ancestor. To examine this problem it will be necessary to reconsider the concept of species from the dynamic point of view, to see by means of what mechanisms differentiation functions among species.

2. MECHANISMS WHICH PRODUCE SEPARATION BETWEEN SPECIES

Sympatric species are reproductively isolated. Discontinuity among the various species occurs in several ways, and knowledge of the means of achieving such discontinuity offers an explanation of the origin of species.

Interspecific sterility is the best documented of this sort of barrier-forming mechanism. Nevertheless, distinct species ("bonae species") exist which, although living in the same area without cross-breeding in their natural state, are found to be capable of mating in captivity. Moreover, these unions produce offspring that are undoubtedly fertile for many generations. Such a barrier is not easily subject to experimentation, nor is it absolute: very often hybrids (for the most part sterile) occur not only between different species but also between different families and genera. Therefore, in nature there must exist mechanisms, other than the barrier of sterility which prevent the mating of members of distinct species.

Seeking to schematize these isolating mechanisms we may begin by dividing them into "extrinsic" and "intrinsic" means. By extrinsic mechanisms we mean all those that are independent of the individual or the species but which have repercussions for the individuals or populations of a species, occasioning, with time, modifications that permit the separation of two populations at the specific level. Geographical isolation is the chief of these mechanisms; over a period of time it facilitates the establishment of mechanisms of genetic isolation.

By intrinsic mechanisms are meant, on the contrary, those which in some way are connected with the individual and with the population. These function either by preventing the individuals from coming in contact during periods of sexual activity or by creating physiological and anatomical barriers so that even though the members of the two populations come in contact they do not mate.

An example of the first mechanism of isolation is that which is found among the various species of *Cercopithecus*, which although living in the same area and in the same trees, are unlikely to meet because they live at different levels (difference of niche). As for the second, it is enough to establish seasonal differences in the sexual activities of different populations even of the same species, or to originate diverse forms of behaviour (courting by the male, specific reaction of the female) to produce isolating mechanisms between individuals living in the same area. Visual, auditory, olfactory and gustatory stimuli are mechanisms which equally serve to determine or maintain isolation. Another mechanism of isolation may be the impossibility of copulation due to major differences in the genital apparatus as is the case in *Macaca arctoides* in respect to the other species of *Macaca* (Fooden, 1967).

Even if a cross should occur, because ecological and ethological barriers do not exist, it may be unfruitful for other reasons which prevent the formation of viable progeny. It may happen, in fact, that the sperm provokes an antigenic type of reaction in the female genital tract and is immobilized and destroyed, or the sperm may reach the ovum but not succeed in penetrating it, or it may penetrate but not produce an effective fertilization, or the ovum may be fertilized but the zygote not develop properly and the embryo abort. Even when all these barriers are overcome and a viable hybrid reaches maturity it is very often found to be sterile. If it is fertile or even partially so, only with difficulty will it succeed in maintaining itself for several generations in a state of nature.

Each of these types of isolation is independent of the others; however, generally if one does not function another begins to operate and the total isolation between two species is usually determined by diverse mechanisms.

3. Pathways to Speciation: the Principal Types of Speciation

What we have briefly described are the mechanisms by which two populations belonging to different species are kept isolated. We have also implicitly touched on some of the means by which populations and species tend to become diversified. We shall now consider in more detail the methods by which two populations can become diversified to the point of constituting two distinct species and which, among speciations of this type, are of most interest in the evolution of the primates and the origin of our species.

Where mutations with a positive value for successive and continuing selections appear in the genetic complex of a species they may establish themselves in one population of the species and represent, when mechanisms of isolation begin to function, the departure base for a new species (autogenous transformation).

In other instances it may be a variation in the number and morphology of the chromosomes of one or more members of a species that may, if of a positive adaptive value, establish itself in a population and be the origin of an incompatibility and, therefore, of a new species (immediate speciation by means of cytological mechanisms).

Variations in one or more environmental factors (change of ecological niche) may determine in some individuals of a species, and in their descendants, differences such as to differentiate them from the other members of the original species. Differentiations of this type may also lead to the formation of new species.

The environment, whether in its physical or biotic guise, may advantageously select different genes. That is, a particular speciation may be due to a semi-geographic isolation with a selective pressure stronger than the flow of genes. The analysis of such a mechanism of speciation is rather complex, and the method which is followed to establish the probability that a character is adaptive consists of a search for correlations between the distribution of a certain character and some environmental element. When such correlations exist, it seems permissible to conclude that the morphological and physiological variations depend on the effect of the special environmental conditions, which vary correspondingly. The variations may be either morphological or physiological. Examples of correlations of this kind might be those between the dimensions of the human nose, or in skin-colour to climate, between thalassaemia and malaria, and so on.

The mechanism which has led to the formation of new species more frequently than any other is undoubtedly geographic isolation followed by the acquisition of barriers to hybridization. This is also the type of speciation which has most influenced the multiplication of different species of primates. There are a number of proofs that speciation is, in fact, brought about by geographic isolation.

In the first place the appearance of reproductive isolation and of differences of ecological requirements in different populations of the same species leads to the formation of two distinct species.

The second proof of geographic speciation consists of the fact that each level of speciation is initiated with populations that are little differentiated, difficult to distinguish, and at the most to be considered as sub-species. If these populations remain geographically distinct they continue to differentiate and may with time reach the level of species. In a second phase, it is possible for the two daughter species, developed from the one original species, to overlap geographically.

The third proof of geographic speciation lies in the existence, in nature, of populations which may still be subspecies or already be species, and about which the taxonomist is unable to decide.

The fourth proof is given by the so-called superspecies, that is groups of related species characteristic of a determinate geographical region.

The fifth proof of geographical speciation is what is termed a double invasion. This occurs when a few individuals invade an island and give rise to a species. If at a later date a second invasion takes place, and in the meantime the first group of colonizers has reached the level of a species, the two groups will live together side-by-side without any possibility of crossbreeding.

The final proof is commonly named "circular overlap". This is the situation when a chain of subspecies is linked by a gradual series of changes. The two terminal populations, although inhabiting the same territory, are reproductively isolated even though connected by an unbroken chain of intermediate populations mutually hybridizing.

It is evident that in a chain of populations in which mechanisms of isolation exist only between the terminal populations, the disappearance of the intermediate links will help in originating distinct species. This is true even if the disappearance is brought about by different causes.

Geographical speciation is thus one of the most important and most effective manifestations in the evolution of taxonomic units and, therefore, in the development of phyletic lines, and, as we have said, also that which has most influenced the evolution of the primates.

4. DYNAMIC DEFINITION OF THE CONCEPT OF SPECIES

From the preceding it is easy to come to the conclusion that a species does not consist of a static taxonomic entity, but because of its biological character, must be defined in an essentially dynamic sense. A tendency of this type is especially evident in the definition of species proposed by Dobzhansky:

"A species is that stage in the evolutionary process in which a group of completely or partially inter-fertile individuals separates into two or more distinct groups, physiologically incapable of reproducing *inter se.*"

Such a definition reconfirms the biological criterion in the definition of a species, and also introduces an evolutionary one.

In the next chapters we will see how this evolutionary vision of the concept of species can be applied to the diverse primate species.

GENERAL REFERENCES

Fooden, J. (1967). Complementary specialization of male and female reproductive structures in the Bear Macaque *Macaca arctoides. Nature, Lond.* **214.** 939–941.

Grant, V. (1963). "The Origin of Adaptations", Columbia University Press, New York and London.

Mayr, E. (1963). "Animal Species and Evolution", The Belknap Press of Harvard University Press, Cambridge, Mass.

Rensch, B. (1959). "Evolution Above the Species Level", Methuen & Co., London.

Simpson, G. G. (1965). "The Geography of Evolution", Chilton Books Publ. Toronto Ambassador Books, Philadelphia and New York.

Tax, S. and Callender, C. (eds) (1960). "Evolution After Darwin", Vols 1–3, University of Chicago Press, Chicago and London.

Washburn, S. L. (ed.) (1963). "Classification and Human Evolution", Methuen & Co./Wenner-Gren Foundation for Anthropological Research, London.

7

Comparison of the Chemical Compositions of Single Molecules in Primates

1. GENERAL

In the preceding chapters we have described the mechanisms which lead to the differentiation of individuals and eventually to that of species. We shall now examine some examples of differentiation which have occurred among the various species of primates. These examples will be considered at the level of the qualitative and quantitative composition of single molecules; the differences between various genes; the products of the interaction of a number of genes, and finally the differences found in the organization of the structures which carry the hereditary characters: the chromosomes.

In this chapter we shall be concerned with the differences in the chemical composition (qualitative and quantitative) of single protein molecules. For the sake of simplicity, and, in order to have more details available, we shall consider haemoglobin exclusively, introducing the subject with a brief account of the principal structural characteristics of a protein.

2. STRUCTURAL DIMENSIONS AND PRINCIPAL CHARACTERISTICS OF A PROTEIN

Proteins are organic substances having very high molecular weights (generally not lower than 15,000) derived from condensation of different amino acids; that is, with the elimination of a molecule of H_2O between the carboxyl group (—COOH) of one amino acid and the amino group (—NH_2) of the next. Proteins may be *simple* if they consist only of a succession of amino acids (united into molecular complexes by the elimination of ammonia between the amino group of one amino acid and the carboxyl group of the next) or *conjugated* if they arise from the union of a simple protein with a non-protein group.

In every protein, different levels of structure called primary, secondary, tertiary and quaternary are recognized. The *primary* structure of a protein is the succession of the amino acids, or fundamental units. The *secondary* structure refers to the spatial organization of the single amino acids and of their links in forming flat rings, as in the case of two cysteine groups linked by an –S–S– bond through the loss of H. The *tertiary* structure is the spatial organization of a chain of amino acids. The *quaternary* structure is the assemblage of subunits of individual polypeptide chains to form the final molecule.

Having amphoteric properties, like any colloid, one protein can be separated from others and from other components of biological liquids by electrophoretic techniques. The basic units of a protein are amino acids, and the number of amino acids which make up a protein chain normally does not exceed 20–25. It is by means of different successions of amino acids that it is possible to have a practically unlimited number of proteins. If the amino acids are compared to the letters of the alphabet, the polypeptides correspond to words and the molecules to sentences. Just as it is possible to express an enormous number of concepts using the twenty-six letters of the alphabet, so with the amino acids an almost unlimited number of proteins may be formed. Each amino acid is characterized by a specific reaction with determined substances or by a particular molecular weight.

An excellent method for distinguishing the various amino acids in a mixture is that of two-dimensional chromatography. In the study of the structure of the proteins, however, it is necessary to know the order in which the amino acids are linked together to form the molecule. To determine the order in which the amino acids of the polypeptide chain of a protein molecule are arranged, the procedure is as follows. The terminal amino acid is marked, for example, by having the amino group react with 2, 4-dinitrofluorobenzene (FDNP). A yellow-coloured compound is thus formed, in which the bond created is more resistant to hydrolysis than the peptidic one (derived DNP). If the protein so treated undergoes a partial hydrolysis, derived dinitrophenols of the terminal amino acid and of the polypeptides of two, three, four . . . amino acids are formed. In a second stage the amino acids which make up the polypeptides are separated by chromatographic methods and identified. For example:

DNP-phenylalanine
DNP-phenylalanine + valine
DNP-phenylalanine + valine + aspartic acid
DNP-phenylalanine + valine + aspartic acid + glutamic acid.

This means that valine is linked to phenylalanine (the terminal amino acid), and this is in turn linked to aspartic acid, which is linked to glutamic acid, and so on. That is, the sequence of amino acids in the pollypeptide chain is exactly as follows:

phenylalanine-valine-aspartic acid-glutamic acid. . . .

Other important data has been supplied about the shape of the protein molecule by X-ray examination.

3. THE HAEMOGLOBINS

Haemoglobin is the respiratory pigment of vertebrates and constitutes almost the total amount of soluble protein in the erythrocytes (about 95%). Haemoglobin belongs to the class of conjugated proteins. It consists of one part—a true protein (globin) which is linked to a second—a prosthetic group (porphyrin) making up about 4% of the molecule (haem or haematin). The porphyrins are cyclic compounds having a large ring made up of 4 pyrrole rings connected by methionine $-CH=$, double bond groups, and having the property of combining with iron and other metals. When the pyrrole rings combine with iron it lies in the centre of the ring, linked to the nitrogen atoms of the pyrrole groups. The haem group so formed is responsible for the red colour of haemoglobin. Both Fe^{2+} and Fe^{3+} ions can link up with porphyrin to form the haem groups, which can therefore exist in both the oxidized (ferric-haem) and the reduced (ferrous-haem) forms. Of these two forms only the reduced one is able to combine reversibly with molecular oxygen.

The haem group is extremely important from the point of view of physiology, but as it is found to have been very stable throughout the course of evolution it does not lend itself to genetic studies: in fact it does not display variations within the primate group. It is therefore of no interest as regards our objective.

Since the content of iron in the haemoglobin of a human adult is about 0·33%, (accepting that each molecule of haemoglobin has a single atom of iron,) it is possible to calculate a molecular weight of about 17,000 for it. Measurements of the sedimentation rate and the osmotic pressure, however, demonstrate a molecular weight of the order of 68,000, which corresponds to a content of four iron atoms, that is, four ferrous-haem groups linked to as many globin chains. An exception to this is the haemoglobin of the lamprey, a primitive chordate, which has a molecular weight of about 17,000 and in this resembles the myoglobin of the higher vertebrates.

4. PRINCIPAL TYPES OF HAEMOGLOBIN IN MAN

In man a number of variants of haemoglobin are known, yet they all correspond to a single basic model common to all the higher vertebrates; that is the union of four subunits each consisting of a polypeptide chain linked to a prosthetic haem group (Figs 7.1 and 7.2).

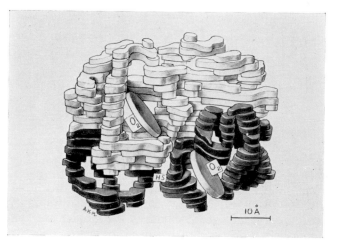

FIG. 7.1 Schematic representation of the molecule of human haemoglobin. The symmetrical disposition of the double α- and β-haemoglobin chains and the models of haemoglobin chains (courtesy of Dr M. Perutz).

The four protein sub-globins of a normal human adult all contain the same terminal amino acid: *valine*. Determination of the penultimate amino acid however distinguishes two varieties: one, called the β chain ends with the succession *leucine, histidine, valine*, while the other, the α chain, ends with the sequence *leucine, valine*.

Foetal haemoglobin is also characterized by four subunits, but they are different from those of the adult: two of them terminate with the amino acid *valine*, but the other two with *glycine*. Those which terminate with *glycine* are characterized by the terminal sequence *Phe-His-Gly*, and are called γ chains; those which terminate with *valine* present terminal sequences of the *Leu-Val* type and are therefore very probably chains very similar if not identical to those of the adult.

At birth foetal haemoglobin is present in the blood of the new-born infant

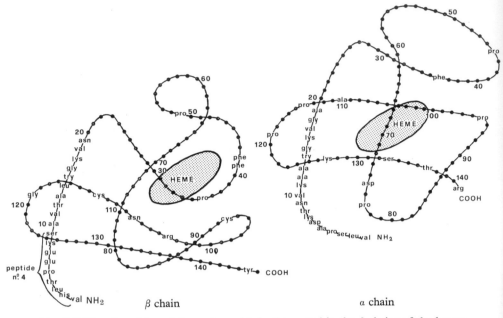

Fig. 7.2 The disposition of the amino acids in the α- and in the β-chains of the haemo-globins (asn = Asp–NH₂).

in the proportion of 60–70%, and is completely substituted by the adult type towards the sixth or seventh month of life. Under normal conditions therefore the γ chain is destined to disappear. In fact in the adult it persists only in individuals affected by a hereditary pathological condition: thalas-saemia. The γ chain differs not only chemically but also physiologically from the other two: foetal haemoglobin has a greater affinity for oxygen than has the haemoglobin of adults.

Another variant of human haemoglobin is the so-called haemoglobin A_2. This type of haemoglobin can be distinguished from the normal only by means of very sensitive electrophoretic techniques. It is formed by two α chains and of two chains which differ from the others described above and are called δ chains. All the variants of the normal human haemoglobins contain 2α chains. Other haemoglobins which differ from the preceding but are found only in pathological conditions (leukaemia) are haemoglobin H, consisting of four β chains, and "Bart's" haemoglobin consisting of four γ chains.

The four types of chain differ among themselves to varying degrees as far as their amino acid composition is concerned: α and β differ in only 21 of the total of 140 amino acids (141 in α and 146 in β), β and γ in 23, β and δ in 10.

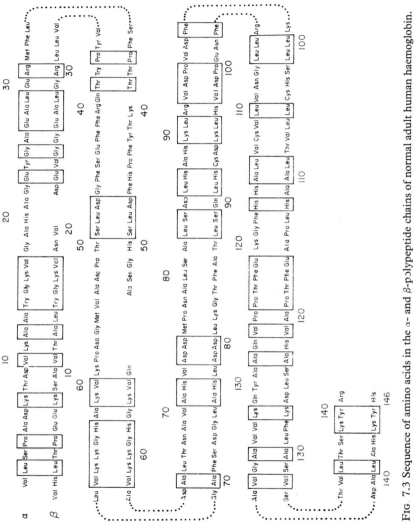

FIG. 7.3 Sequence of amino acids in the α- and β-polypeptide chains of normal adult human haemoglobin.

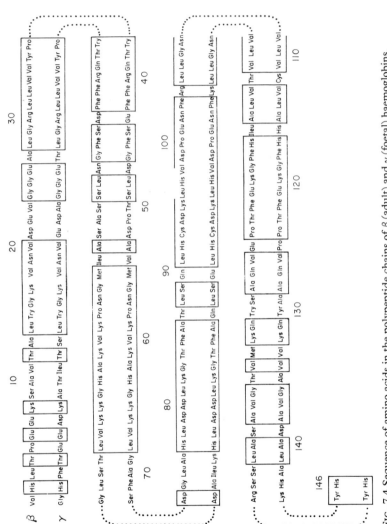

FIG. 7.4 Sequence of amino acids in the polypeptide chains of β (adult) and γ (foetal) haemoglobins.

5. COMPARISON OF THE HUMAN GLOBIN CHAINS

Considering the succession of amino acids in the α and β chains, it may be noted that, notwithstanding the differences in length, large portions of the two chains show a notable mutual similarity. They are equal for approximately 42% of their lengths, containing 60 identical amino acids in the same relative positions. This suggests that the two chains were identical when they originated, and that they only diversified later through deletions and mutations in the information-carrying DNA. In support of this hypothesis it may be observed that the α chain differs from the β chain more than the γ differs from the δ chain. It would seem therefore that the α chain must have undergone the greatest degree of differentiation.

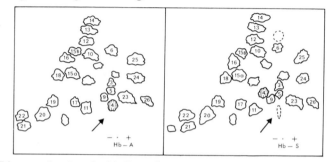

FIG. 7.5 Diagram showing the bi-dimensional electrophoretic profile (fingerprint) of the tryptic polypeptides derived from HbA (Zuckerkandl *et al.*, 1960).

For a comparative study of the polypeptide chains of haemoglobin it is easier and faster in the initial stages to compare their "fingerprints": two dimensional electrophoretic maps of the tryptic peptides, derived from the enzymatic hydrolysis of the globin, by trypsin than to attempt the very complicated analysis of the sequence of the amino acids.

Tryptic hydrolysis liberates about 14 or 15 peptide fragments from each chain. Of these only one (a pentapeptide) occupies the same position on the two-dimensional maps (fingerprints) of the α, β and γ chains. It is therefore possible to affirm that the structural differences between these three chains are distributed along the entire length of the polypeptide. On the other hand, if the β and γ chains are compared, it is seen that 8 peptides occupy equivalent or at least very similar positions in the "fingerprints".

This indicates a remarkable homology between the two chains. The β and γ chains differ from one another in only four peptides and this is truly a minimal difference concerning only four main single amino acid substitutions. The pathological variations in haemoglobins are sometimes due to the substitution of only one amino acid. This occurs in the HbS mutation in which the valine (position 6 in peptide 4) is substituted by glutamic acid, and also occurs in the HbG mutation in which the glutamic acid (position 7 in peptide 4) is substituted by glycine.

6. The Haemoglobin of the Primates

In recent years many hypotheses have been propounded and many research projects undertaken to determine the steps in the evolutionary process which have led to the formation of the genes which preside over the synthesis of the polypeptide chains of normal human haemoglobin. The most interesting results have been produced by the systematic study of amino-acid sequences in the polypeptide chains of the haemoglobin of various primate species.

The structural homologies that can be ascertained by investigations of this sort help to establish what the evolutionary stages have been, or at least the direction followed by the genes which control haemoglobin.

Research carried out on the haemoglobin of primate erythrocytes subjected to tryptic hydrolysis and then studied by means of two-dimensional electrophoresis or chromatography, has produced the same results nearly every time. In some cases, however, the sensitivity of the methods of investigation have been such that only with one method of the two has it been possible to bring differences to light. This, for example, is the case in the comparison of human haemoglobin with that of *Hylobates lar*. In this instance the examination of the tryptic peptides by two-dimensional electrophoresis did not reveal any difference between the tryptic peptides of the human α and β chains and those of the corresponding chains of the gibbon, while chromatography on ion exchange resins gave clear evidence of a difference at the level of the ninth peptide of the β–chain (β Tp IX).

The differences in the composition of the amino acids of analogous chains of prosimians and man are greater than those found between monkeys and man, and these differences steadily diminish as species more closely related to us are compared. The greatest difference is the presence of 1–3 residues of iso-leucine, which is completely lacking in man, in both the α and β chains of the haemoglobin of the lemurs. The amino-acid composition of some pure peptides obtained from the β chain of a number of primates is set out in Table 7.1.

Although the sequence of the peptides has not yet been recorded with certainty, their succession has been fixed by analogy with the human β and γ chains. Only rarely are there three or four amino acids in each peptide that are not present in the analogous peptide of the human chain, and the number is nearly always similar.

Therefore, in attempting to develop a succession of sequences those amino acids which appear in both the human and primate peptides have been placed in the position which they are found to occupy in the peptide of the human β chain, for which the sequence is well known. In order to assign a

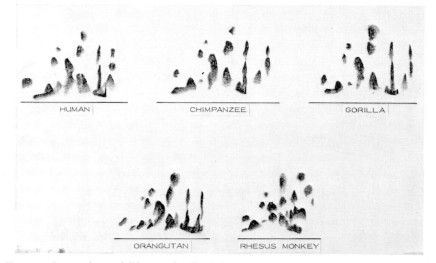

FIG. 7.6 Comparison of "fingerprints" of the haemoglobin of different primates (from Zuckerkandl *et al.*, 1960).

position to the other amino acids, as if they were the tesserae of a mosaic, account has been taken of the fact that many of the amino acids present in the haemoglobin of the primates in question and not found at the level of the human β chain, are, however, present in the human γ chain. In this way it has been possible to establish, even if not with absolute certainty, the greater part of the sequence of amino acids in the haemoglobin of the non-human primates. As an example of this type of reconstruction there follows the proposed sequence of amino acids of the first tryptic peptide of the β chain (β T PI) in *Galago crassicaudatus*:

		1	2	3	4	5	6	7	8
Tp–1	*Homo* (β)	Val	His	Leu	Thr	Pro	Glu	Glu	Lys
Tp–1	*Galago*	Val	His	Phe	Thr	Pro	Gly	Asp	Lys
Tp–1	*Homo* (γ)	Gly	His	Phe	Thr	Glu	Glu	Asp	Lys

The 5 amino acids which appear both in *Galago* and in *Homo* in the β chain have, by analogy, been assigned to the positions which they occupy in human chain: 1, 2, 4, 5, 8. Phenylalanine and aspartic acid have been placed in positions 3 and 7 which they occupy in the human γ chain: and the last remaining amino acid, valine, has been given the still vacant sixth position.

The number of differences found between primate and human haemoglobin, as we have said, seems to be directly proportional to the phylogenetic distance under consideration. The nearer the creatures are on the scale

TABLE 7.1. Amino acid composition of some peptides of the β and γ chains in some non-human primates and man. The italics indicate those amino acid residues different from corresponding residues of the β-chain in human haemoglobin (according to Hill et al., 1963).

	1	5	(βTp-1)	18	20	25	(βTp-III)

β Chain of man val-his-leu-thr-pro-glu-glu-lys.....val-asg-val-asp-glu-val-gly-gly-glu-ala-leu-gly-arg

Hylobates lar (val,his,leu,thr,pro,glu,glu,lys).....(val,asp,val,asp,glu,val,gly,gly,glu,ala,leu,gly,arg)

Papio doguera .....(val,asp,val,asp,glu,val,gly,gly,glu,ala,leu,gly,arg)

Galago crassicaudatus (val,his,*phe*,thr,pro,*gly*,*asp*,lys).....(val,asp,val,*glu*,glu,val,gly,gly,glu,ala,leu,gly,arg)

Perodicticus potto (val,his,leu,thr,*glu*,*gly*,*asp*,lys).....(val,asp,val,asp,glu,val,gly,gly,glu,ala,leu,gly,arg)

Propithecus verreauxi .....(val,asp,val,*glu*,*asp*,*ala*,gly,gly,glu,*thr*,leu,gly,arg)

Lemur variegatus .....(val,asp,val,*glu*,*lys*)

L. catta .....(val,asp,val,*glu*,*lys*,val,gly,gly,glu,*thr*,leu,gly,arg)

L. fulvus .....(val,asp,val,*glu*,*lys*,val,gly,gly,glu,ala,leu,gly,arg)

(γTp-1) (γTp-III)

	31	35	(βTpIV)	45	50

γ Chain of man val-his-*phe*-thr-*glu*-*asp*-lys.....val-asg-val-*glu*-*asp*-*ala*-gly-gly-glu-*thr*-leu-gly-arg-

β Chain of man leu-leu-val-val-tyr-pro-try-thr-glu-arg-phe-phe-glu-ser-phe-gly-asp-leu-ser-pro-

Hylobates lar (leu,leu,val,val,tyr,pro,try,thr,glu,arg)(phe,phe,glu,ser,phe,gly,asp,leu,ser,thr,pro,

Papio doguera (leu,leu,val,val,tyr,pro,try,thr,glu,arg)(phe,phe,glu,ser,*leu*,gly,asp,leu,ser,*glu*,thr,pro,

Galago crassicaudatus (leu,leu,val,val,tyr,pro,try,thr,glu,arg)(phe,phe,glu,ser,*leu*,gly,asp,leu,ser,thr,pro,

Perodicticus potto (leu,leu,val,val,tyr,pro,try,thr,glu,arg)

Propithecus verreauxi (leu,leu,val,val,tyr,pro,try,thr,glu,arg)(phe,phe,gly,ser,phe,gly,asp,leu,ser,*ser*,pro,

Lemur variegatus (leu,leu,val,val,tyr,pro,try,thr,glu,arg)(phe,phe,glu,ser,phe,gly,asp,leu,ser,*ser*,pro,

L. catta (leu,leu,val,val,tyr,pro,try,thr,glu,arg)

L. fulvus (leu,leu,val,val,tyr,pro,try,thr,glu,arg)

(γTp-IV)

γ Chain of man   leu-leu-val-val-tyr-pro-try-glu-arg-phe-phe-asp-ser-phe-gly-asg-leu-ser-ser-ala-

     55   (βTp-V)   (βTp-VI)   (βTp-VII)   70

β Chain of man   asp-ala-val-met-gly-asg-pro-lys-val-lys-ala-his-gly-lys-lys-val-leu-gly-ala-phe-ser-

Hylobates lar   asp,ala,val,met,gly,asp,pro,lys⟩⟨val,lys⟩⟨ala,his,gly,lys⟩   ⟨val,leu,gly,ala,phe,ser,

Papio doguera   asp,ala,val,met,gly,asp,pro,lys⟩   ⟨val,leu,gly,ala,phe,ser,

Galago crassicaudatus   glu,ala,val,met,gly,asp,pro,lys⟩⟨val,lys⟩⟨ala,his,gly,lys⟩   ⟨val,leu,glu,ser,phe,ser,

Perodicticus potto   ⟨val,lys⟩   ⟨val,leu,thr,ser,phe,gly,

Propithecus verreauxi   ser,ala,ileu,met,gly,asp,pro,lys⟩⟨val,lys⟩   ⟨val,leu,gly,ser,phe,ser,

Lemur variegatus   ser,ala,ileu,met,gly,asp,pro,lys⟩⟨val,lys⟩   ⟨val,leu,thr,ser,phe,gly,

L. catta   ⟨val,lys⟩⟨ala,his,gly,lys⟩

L. fulvus   ⟨val,lys⟩

     (γTp-V)   (γTp-VI)   (γTp-VII)

γ Chain of man   ser-ala-ileu-met-gly-asp-pro-lys-val-lys-ala-his-gly-lys-   va-leu-thr-ser-leu-gly-

     75   80 (βTp-IX)

β Chain of man⟩
Hylobates lar   asp-gly-leu-ala-his-leu-asp-asp-leu-lys

Papio doguera   asp,glu,leu,ala,his,leu,asp,asp,leu,lys⟩

Galago crassicaudctus   asp,asp,leu,ala,his,leu,asp,asp,leu,lys⟩

Perodicticus potto   asp,ala,val,ala,his,leu,asp,asp,leu,lys⟩

Propithecus verreauxi   asp,ala,leu,glu,his,leu,asp,asp,leu,lys⟩

Lemur variegatus   glu,glu,thr,pro,his,leu,asp,asp,leu,lys⟩

     (γTp-IX)   (γTp-X)

γ Chain of man   asp-ala-ileu-lys-his-leu-asp-asp-leu-lys

of evolution the smaller are the differences at the level of the composition of the polypeptide chain of the globin. The use of amino acid sequences for working out primate phylogenesis is obviously of great importance. Great progress in this area has been recently achieved, however, the technique requires knowledge both of protein chemistry and systematic biology. Unfortunately, few scientists are trained in both areas and misunderstandings have arisen. Few sequences have so far been worked out, but those that have are interpreted to show that most substitutions are of neutral selective value, and this leads to the conclusion that their accumulation should be time dependent. Thus the length of time that has elapsed since the member of two given species last shared a common ancestor should be proportional to the number of differences in the sequences of homologous proteins found in these species. However this assumption may not be absolutely true and some selective forces and preferential substitution probably worked also in the differentiation of the primate haemoglobin chains.

A summary of the substitution in primate α and β chains is shown in Tables 7.2 and 7.3. The sequence of the ancestral chain (precursor) was derived from the data on the primate chains plus the data on other mammalian α chains. Blank spaces in the sequences indicate that the residue is identical to that found in the hypothetical ancestral chain. The sequence positions not shown are invariant in all primate species examined.

In the α chains the anthropoid apes are not mentioned, since their sequences are almost identical to man. *Papio* and *Macaca* differ from each other by at least 13 substitutions in the α chains. However, when compared with human α chain, *Macaca* and *Papio* both share the substitutions at position 68 and 71. Since *Macaca* and *Papio* are closely related these data emphasize the qualitative aspect of sequence comparison and de-emphasize the quantitative one. The location of common mutations determines phylogenesis; the number of substitutions may or may not be related to the time divergence.

Five positions (12, 16, 24, 129, 131) link together *Propithecus* and *Lemur*. The substitution of methionine to leucine in position 76 is the only clear link of the prosimians with the higher primates while the substitution of histidine to leucine in position 113 links all the prosimians together, *Tupaia* included. Position 113 also links prosimians with the insectivores.

The sequence comparison of primate β chains is shown in Table 7.3. It immediately appears that the Prosimians and the Simiae are descended from separate lines, even though few species have been examined.

The prosimians in fact show a completely different pattern in amino acid substitutions from the Old and New World Monkeys and the Apes.

The substitutions at several positions are, however, of interest: glutamine at position 87 is the obvious precursor in the chains of all Primates and the substitution of glutamine by threonine in the Apes and Man is a recent one that seems to link them together. Glutamine in position 125 appears to unite the Old and New World Monkeys. The homologies of position 5 (proline) and 56 (glycine) also unite Old World Monkeys, Apes and Man and differentiate them from the New World Monkeys and Prosimians. Position 56 may, moreover, indicate a relationship between New World Monkeys and Lemurs.

When the composition of the amino acids of the peptides of the β chain of the most primitive primates is examined more carefully, one is struck by the similarity between these and the γ chain characteristic of the human foetus. The first to the ninth peptides of the human β chain contain altogether 83 amino-acid residues of which 20 differ from those of the γ chain. Most of the differences in the amino-acid composition of the β chains of the primates studied may be referred to the 20 positions for which the human β and γ chains are different. Moreover when the composition of one of the peptides of the primates differs from the homologue of the human β chain, the different amino acids are actually those of the human γ chain.

On the basis of these observations it may be concluded that during phylogenesis the polypeptide chains of the haemoglobin of the primates demonstrated varying degrees of stability. The α chain remained more stable than the β chain in the course of the phylogenesis of the Primate Order, and the β chains of the less-evolved primates contain many amino acids that are present in the γ chain of Man.

Notwithstanding the apparent "instability" of the β chain in the course of evolution, it is evident that certain sequences of these chains do not demonstrate major variations: this can be shown from the succession of residual amino acids from the 30th to the 40th positions which is the same in all the chains of the Primates: Leu-Leu-Val-Val-Tyr-Pro-Tyr-Thr-Glu-Arg N. The stability of this sequence in the β chains, suggests that it must be very important for the functioning of the molecule and therefore the mutations which might be registered in this zone have not been confirmed because they are injurious or lethal.

7. The Evolution of the Haemoglobin Molecule

Evolution comes to pass, as we have seen, by means of natural selection working on spontaneous mutations. This assertion implies, first, the existence of a mechanism which ensures the constancy of the genetic information, and, second, the occurrence of spontaneous mutations on which selection can act. The genetic information written in the form of triplets (purine and pyrimidine bases) in the information-carrying DNA, which codifies the succession of amino acids in protein molecules, provides for the

TABLE 7.2. Alignment of primate α chains (from Sullivan 1971).

Position	4	5	8	10	12	15	16	17	19	20	22	23	24	48	53	55	56	57
Precursor	Pro,	Ala,	Ser,	Val,	Ala,	Gly,	Lys,	Ile,	Gly,	His,	Gly,	Glu,	Tyr,	Leu,	Ala,	Val,	Lys,	Ala,
Homo			Thr,					Val,	Ala,									
Macaca								Val,									Gly,	Gly,
Lemur			Thr,					Val,		Glu,			His,		Gly,			
Propithecus	Ala,				Leu,	Thr,	Lys,	Ala,	Ser,				His,					Thr,
Galago			Ile,					Val,			Glu,	Asp,	Met,	Met,				
Tupaia	Gly,	Thr,				Glu,		Ile,		Glu,	Pro,							

Position	67	68	71	73	76	78	96	105	111	113	114	115	116	117	125	129	131
Precursor	Thr,	Lys,	?,	Leu,	Leu,	Gly,	Val,	Leu,	Ser,	Leu,	Pro,	Ala,	Glu,	Phe,	Leu,	Leu,	Ser,
Homo		Asn,	Ala,	Val,	Met,	Asn,			Ala,								
Macaca		Leu,	Gly,	Val,	Met,	Asn,			Ala,								
Lemur			Ser,							His,							
Propithecus	Gly,	Asp,			Met,	Thr,				His,	Ser,						
Galago		Ser,		Val,	Met,	Ser,			Cys,	His,					Leu,	Phe,	Met,
Tupaia	Ser,	Thr,	Gly,			Thr,	Ala,	Ile,	Cys,	His,			Gly,	Asp,			

TABLE 7.3. Alignment of primate β chains (from Sullivan 1971).

Position	1	2	4	5	6	7	8	9	10	12	13	16	17	21	22	33	43	50	52	54	56
Homo	Val,	His,	Thr,	Pro,	Glu,	Glu,	Lys,	Ser,	Ala,	Thr,	Ala,	Gly,	Lys,	Asp,	Glu,	Glu,	Thr,	Asp,	Val,	Gly,	
Gorilla																					
Macaca								Asn,								Leu,	Ser,				
Cebus				Ala,							Thr,						Asp,				
Saimiri				Gly,	Asp,			Ala,			Thr,										
Ateles				Gly,						Ala,											
Saguinus																					
Lemur	Thr,	Leu,	Ser,	Ala,		Leu,	Asp,	Ala,	His,		Thr,	Ser,		Glu,	Lys,			Asn,	Ser,	Ser,	
Propithecus	Thr,	Leu,	Gly,	Asp,	Val,		Ser,	Ala,	Glu,		Ser,	Ala,	Asp,	Glu,	Lys,			Ser,	Ser,	Ile,	Ser,

Position	69	70	73	76	87	104	111	112	113	114	115	116	119	120	121	122	123	125	126	130	135	139
Homo	Gly,	Ala,	Asp,	Ala,	Thr,	Arg,	Val,	Cys,	Val,	Leu,	Ala,	His,	Gly,	Lys,	Glu,	Phe,	Thr,	Pro,	Val,	Tyr,	Ala,	Asn,
Gorilla																						
Macaca			Asn,		Gln,	Lys,									Gln,							
Cebus				Thr,	Gln,	Lys,									Gln,							
Saimiri		Thr,			Gln,										Gln,					Thr,		
Ateles					Gln,										Gln,		Leu,					
Saguinus					Gln,										Gln,							
Lemur	Ser,	Glu,	His,		Gln,	Lys,						His,	Asp,		Ser,		Leu,				Ala,	Phe,
Propithecus	Ser,	Glu,	Asp,		Gln,							His,	Asp,		Ala,			Ser,	Glu,		Phe,	Thr,

first condition, and the second is provided for by a variety of diverse physical and biological factors. Selection has exerted its effects in various ways on different mutations, eliminating the disadvantageous and lethal and adopting the advantageous. Every molecule is in the optimum state which results from interaction between environment and the intrinsic plasticity of the molecule.

No variation is complete in itself insofar as it conditions the existence of differentations beyond itself at various levels of the organism. If, for example, in a particular protein molecule, a tyrosine happens to replace a phenylalanine, a new supplementary peptide results. This peptide, like many others, will require specific enzymes, both for its synthesis and its destruction. If the constitution of the peptide is new to the organism these enzymes may be lacking, and a lethal or injurious effect may result so that it would be difficult for such a mutation to establish itself.

The substitution of an amino acid does not involve only the primary structure of a protein, but can also influence the secondary and tertiary structures. Sometimes it may even produce alterations in its spatial organization (quaternary structure).

The study of the phylogenesis of proteins is carried out by comparing "homologous proteins" (that is, proteins that are functionally equivalent) isolated from organisms situated at different levels in the same phyletic series but if possible close to each other.

Up to the present the function of a protein as, for example, its capacity to link itself reversibly with the oxygen molecule, has always been accepted as a basis of identification, and only recently have the structural homologues been established. However, this does not mean that the existence of different proteins which are structurally homologous but designed for diverse functions must be excluded.

The three variants of haemoglobin which are found in normal human blood—HbA of the adult, foetal HbF and Kunkel's HbA$_2$—represent different combinations of four different polypeptide chains (α, β, γ and δ) the structures of which are genetically determined. In fact it is thought that four loci exist which are responsible for the difference sequences of the amino acids of the four chains. Geneaological research indicates that the structural genes which determine the primary structure of the β and δ chains are strictly linked together, while the genes for the α and γ chains segregate independently. However, there is no reliable test to indicate on which chromosome the four structural loci and their correlated synthesis-controlling genes are located.

The globin chain of human haemoglobin has been compared with analogous protein chains of various vertebrates. The outline of molecular evolution of the genes for the polypeptide chains of haemoglobin (as conceived by Ingram) takes as its point of departure, the time before the appearance of the fish, when it is assumed that the haemoglobin of the blood was identical with the myoglobin of the muscles. Myoglobin, in fact, is an intracellular respiratory pigment, found in both the vertebrates and invertebrates, having a molecular weight of about 17,000. It contains a haem group and seems to be made up of a solitary polypeptide chain. This is indicated by the presence of a single N-terminal amino acid (valine in the horse, glycine in the whale).

It is supposed that in the course of evolution, beginning with a gene for "primitive haemoglobin" presumably containing an α chain very similar to that of existing haemoglobin; a duplication of genes occurred following a translocation. Following this stage each of the two α genes evolved independently, one becoming the gene for present day myoglobin, which has preserved the principal ancestral characteristics, the other becoming the gene for the α chain of adult haemoglobin. Later this gene would have evolved toward the dimerous form α_2. Actually the dimerous condition with a haem-haem interaction has a selective advantage as compared with the monomer, as is shown by the dissociation curve in Fig. 7.7.

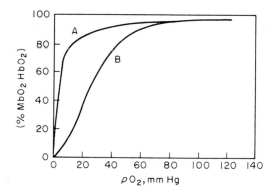

FIG. 7.7 Curve of dissociation of oxygen for myoglobin (A; monomer) and for haemoglobin (B; tetramer).

It is postulated that at this stage the α_2 genes underwent a new duplication, and after this, the two dimers α_2 and γ_2 appeared side by side. At this point the pressure of selection would have led to the stabilization of tetramers with a further genic duplication. A condition of this type is found in fish, where in certain cases a haemoglobin having four polypeptide chains and four haem groups is seen, while in others a haemoglobin formed of a single chain is present, and thus is presumably very similar to the ancestral haemoprotein.

A third translocation with reduplication may be placed at the level of the formation of the gene for the primitive γ chain, which was one of the two genes destined to evolve along parallel lines to originate the gene for the β chain. The formation of the tetramer was already well stabilized at this stage and the new gene produced haemoglobin A (α_2, β_2), the dissociation curve of which shows an optimum adaptation to the needs of the adult human organism. On the other hand, the ancient γ chain which continued

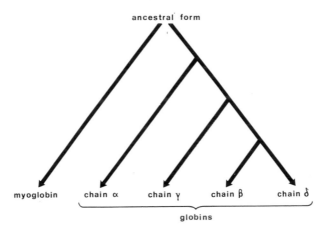

FIG. 7.8 Scheme of the evolution of haemoglobin through genic duplication followed by translocation (according to Ingram's hypothesis).

to be synthesized by one of the two genes resulting from the last duplication and translocation, maintained with the α chain the formation of the tetramer α_2, γ_2 characteristic of foetal haemoglobin.

The theory, according to which the gene for the β chain is the result of a mutation of the γ gene rather than the α gene, is confirmed by the fact that the β and γ chains resemble each other more closely than do the α and β or α and γ chains. At this point we have three independent genes α, β, and γ, which associate in dimerous form.

The formation of the two tetramers $\alpha_2 \beta_2$ and $\alpha_2 \gamma_2$ for the adult and foetal haemoglobins results from the association of these dimers.

A later duplication of the β gene (not in this case followed by translocation) produced the δ gene from the β which are still linked together.

At this point the question arises as to how many mutations were necessary to arrive at the present human gene starting from its primitive ancestor, or between the later and the existing human α gene. It is difficult to answer such a query, and it would perhaps be more reasonable to ask what must have been the minimum number of mutations separating the present-day human α and β chains. Of the 141 amino acids of the α chain 85 vary in sequence from the β chain. This indicates that a minimum of 85 mutations must have intervened, each having brought about the substitutions of single amino acids.

It is generally believed that the evolution of the Vertebrates began 5×10^8 years ago.

If we assume that the average period of a generation among the Vertebrates is five years, this would imply 10^8 generations for the evolution of vertebrate haemoglobin. The frequency with which mutations are usually taken to occur in one generation is 10^{-5} or 10^{-6}, from which it is possible to calculate that from 100 to 1000 mutations have taken place in the history of the evolution of the vertebrate haemoglobin.

On the basis of these data an hypothesis can also be formulated concerning the chronological periods of the appearance of the different genes which have led to the differentiation of the various globin chains. Recently Zuckerkandl and Pauling have established that the common precursor of the α and γ chains must have appeared around 600 million years ago and the precursor of the γ and β chains about 260 million years ago, while that of the β and α chains must be much more recent: 44 million years ago.

Considering that the evolution of the mammals including the primates, has taken place during the last 160 million years, an analysis of the haemoglobins of the primates cannot be expected to throw light on any but the last stages of the evolution of the various haemoglobins. The data set out in the preceding chapter would suggest that the γ chains appeared for the first time only in the most primitive primates. If this is true, it is possible that the time calculated for the derivation of the α and γ chains, from a common precursor (260 million years ago) could be much reduced, and that the evolution of the β chain, from a mutual precursor of the γ type could have occurred within the 160 million years of mammalian evolution.

8. Genetic Distances and Phylogenetic Trees

Other protein molecules besides haemoglobins can be used for studying the amino acid sequence differences which are considered to be due to single gene mutation, and they are, therefore, useful for calculating genetic distances among different species. This data, moreover, can provide us with a sort of evolutionary clock. In fact, at first approximation, the amount of divergence between two species can be measured by counting the ratio of the number of homologues that have a different amino acid.

As stated before, the use of protein sequence data as taxonomic characters

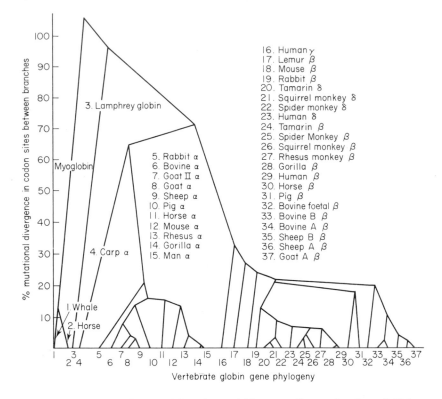

Fig. 7.9 Phylogenetic tree of Vertebrate globins according to Goodman (1971).

as opposed to many anatomical characters can help one to arrive at a cladistic phylogenetic tree without taking into account the pleiotropic and polygenic characters which, as we will see, are more difficult to analyse. Necessity demands, however, that more than one protein is used, and the cladograms obtained are compared.

Various attempts have been made previously to reconstruct the evolutionary pattern of different primate species using the amino acid sequences in different protein molecules. Besides haemoglobin, cytochrome *c*, insulin and fibrinogen have been fairly extensively used for tracing phylogenetic trees for the entire group of Mammals.

The organization of these phylogenetic trees, however, deserves particularly complex procedures. A full discussion of the logical and statistical problems of the tree-building procedure has been recently discussed by Cavalli-Sforza and Edwards (1967).

The procedure most generally used at present is the one devised by

Margoliash and Smith. It consists in calculating the "minimal mutational distance" between all pairs of the species considered. This means calculating the minimum number of nucleotide changes required to account for the particular amino acid replacements. These mutational distance values are processed by a special computer programme. The computer tries all the possible associations and divisions for all the species for which data are available and automatically builds a tree on the basis of a previous programme.

One of these phylogenetic trees which is comprehensive and takes into consideration the different globin chains in Vertebrates is presented in Fig. 7.9.

GENERAL REFERENCES

Anfinsen, C. B. (1959). "The Molecular Basis of Evolution", J. Wiley and Sons, New York.

Baglioni, C. (1962). Correlation between genetics and chemistry of human haemoglobins. In "Progress in Molecular Genetics" (H. Taylor, ed.), Academic Press, New York and London.

Cavalli-Sforza, L. L. and Edwards, A. W. F. (1967). Ann. J. hum Genet. 19, 233.

Florkin, M. (1966). "Aspects moleculaires de l'adaptation et de la phylogenie", Masson and Cie, Paris.

Goodman, M., Koen, A. L., Barnabas, J. and Moore, G. W. (1971). Evolving primate genes and proteins. In "Comparative Genetics in Monkeys, Apes and Man" (B. Chiarelli, ed.) Academic Press, London and New York.

Hill, R. L., Buettner-Janusch, J. and Buettner-Janusch, V. (1963). Evolution of hemoglobin in Primates. Proc. natn. Acad. Sci. U.S.A. 50, 885.

Huens, E. R. et al. (1964). Developmental hemoglobin anomalies in a chromosomal triplication: D_1 trisomy syndrome. Proc. natn. Acad. Sci. U.S.A. 51, 89.

Ingram, V. M. (1961). Gene evolution and the hemoglobins, Nature Lond. 189, 704.

Ingram, V. M. (1961). "Hemoglobin and its Abnormalities", Charles C. Thomas, Springfield, Illinois.

Ingram, V. M. (1963). "The Hemoglobins in Genetics and Evolution", Columbia University Press, New York and London.

Söderqvist T. and Blömbäck B. (1971). Fibrinopen structure and evolution. Naturwissenschaften 58, 16–23.

Sullivan, B. (1971). Comparison of the Hemoglobins in Non-human Primates and their importance in the Study of Human Hemoglobins. In "Comparative Genetics of Monkeys, Apes and Man" (B. Chiarelli, ed.) Academic Press, London and New York.

Zuckerkandl, E., Jones, R. T. and Pauling, L. (1960). A comparison of animal haemoglobins by tryptic peptide pattern analysis. Proc. natn. Acad. Sci. U.S.A. 46, 1349.

Zuckerkandl, E. and Pauling, L. (1968). "Structural Chemistry and Molecular Biology", D. Van Nostrand Co., Princeton, N.J.

NOTE ADDED IN PROOF

The most recent discoveries in the field of molecular biology and evolution of Primates are collected in the special issue of the *Journal of Human Evolution* Vol. **6** (1972) organized by Morris Goodman. This issue is suggested to the readers interested in this field.

8

Comparison of Single Hereditary Characters of Man with Those of Other Primates

1. COMPARATIVE GENETICS OF THE PRIMATES

Similarities and differences can be distinguished not only by comparison at the molecular level of the chemical composition of individual proteins, but also by studying the manifestations of single genes and their frequencies within the ambit of the various species. Comparative study at the level of single hereditary characters in the various species of primates is interesting not only because it makes possible the reconstruction of the pattern of evolution followed in nature for the appearance and establishment of single hereditary characters, but also because it affords an opportunity for singling out analogous hereditary characters in animals that certainly resemble man more closely than does *Drosophila* or the mouse. If the attention of human geneticists is directed towards these analogous characters, the behaviour of those of our species will be better understood. In many cases it is probable that the differences between species represent an extension of intraspecific variations. Certainly the human geneticist will be better able to appreciate how hereditary characters vary within the limits of our species when he knows how they vary between the species most closely related to us. This is not to say absolutely that the presence of the same gene in two different species indicates a common phylogenetic origin. Part of the genic complex may be maintained almost intact in two species which have been distinct for a long time, while other closely related species or even two populations of the same species may diverge through the loss or the extensive modification of one or more genes.

The science of comparative genetics of the primates is a very new one. Useful data are still extremely scarce and it is premature to think of finding association-groups and even more premature to hope to make genetic maps for the different species of primates, when only in the last years have we glimpsed the possibility of doing this for man. Nevertheless the data are promising. We shall give examples of some of the best and most extensively known characters.

2. Blood Groups

The substances which determine human blood groups are mucoproteins or mucopolysaccharides; that is, large protein molecules to which various molecules of sugars are attached. These substances found in the blood are also called antigens. An antigen may be defined as a large protein molecule (at least *usually* a protein) that stimulates the production of an antibody or reacts with an antibody already present in the serum. Antibodies are proteins of large dimensions which form part of the γ- globulin fraction of the serum. An organism generally responds to the presence of an extraneous substance by producing an antibody against it. The antibodies in which we are interested are those which agglutinate the red corpuscles; they carry specific antigenic groups on their surfaces. Almost all the group-specific antigens of the blood are present at birth, and the antibodies against some of them are also present in the circulating blood from that time. They are, therefore, predetermined by heredity in each individual, and are characteristic not only of the cells circulating in the blood but also of all the tissues of the organism. In 1900 Landsteiner, mixing red corpuscles with the serum of different individuals, demonstrated that the red corpuscles of some individuals are agglutinated by the serum of others. He was thus able to distinguish the blood types of individuals whom we now classify in groups A, B, O and AB. It need not be said how important this discovery has been to medicine, for millions of people have survived serious wounds or major surgery thanks to blood transfusions. In addition the discovery has been of the greatest importance in genetics and anthropology.

These characters are absolutely hereditary and follow the laws of Mendel so well that they can be used as a test for excluding paternity. The correlation between the phenotypes A, B, O and AB and their genotypes is given in Table 8.1.

Other systems of blood groups have been determined by injecting the red blood cells of human beings into animals such as the rabbit and *Macaca rhesus.* These animals synthesize specific antibodies just as we immunize ourselves against poliomyelitis when we are vaccinated. By the antisera produced by the plasma of the immunized animals mixing with red corpuscles of various human beings, it is possible to decide which individuals possess a particular antigen and which do not. The red blood cells that contain the antigen are agglutinated by the anti-serum containing a specific antibody, while those which do not contain it are not.

TABLE 8.1. Correspondence between the phenotypes A, B, O and AB and their genotypes.

Genotype	Phenotype
OO	O
AO	A
AA	A
BO	B
BB	B
AB	AB

Another method by which it is possible to distinguish blood groups is that of using the serum of people who have received repeated blood transfusions. These individuals are normally immunized against the antigens contained in the red corpuscles of the donor or donors. If we mix the red blood cells of the donor with the serum of the recipient, very often specific antibodies against the red corpuscles of the donor, which agglutinate his blood, are found to be present. Another situation in which discovery of new systems of blood groups is possible is the presence in the newborn of haemolytic disorders. In fact, antibodies not only agglutinate the red corpuscles, but in some cases may also cause them to rupture (haemolysis). In the newborn the condition called "haemolytic icterus" occurs when the antibodies which may be present in the mother's serum succeed in passing across the placenta and enter the foetal circulation to attack and destroy the red blood cells of the foetus. In many cases this haemolysis is caused by an antibody normally circulating in the mother's blood, such as anti-A or anti-B. In other cases the antibody may be synthesized by the antibody producing tissues of the mother under the stimulus of antigens present on the erythrocytes of the foetus which by chance may pass into the maternal circulatory system across the placenta.

By these techniques or expedients, numerous systems of blood groups in man have been identified. The principal ones are listed in Table 8.2.

TABLE 8.2. The principal human blood group systems. (From Buettner-Janusch, 1966).

System	Antigen	Antibodies	Phenotype	Genotype
ABO	A, B, [AB]	anti-A anti-B	O, A, B, AB	OO, AA, BB, AB, AO, BO
Lewis	Le^a, Le^b	anti-Le^a anti-Le^b	Le (a+b−), Le (a−b+), Le (a−b−)	Le^aLe^a, Le^bLe^b, $LeLe$
Rh				
MNSs	M, N, S, s	anti-M anti-N anti-S anti-s	M, N, MN, S, s, Ss	MS/MS, MS/Ms, Ms/Ms, MS/NS, MS/Ns, Ms/NS, Ms/Ns, NS/NS, NS/Ns, Ns/Ns
P	P_1, P_2	anti-P_1 anti-$P+P_1$ anti-P	P_1, P_2, p	P_1P_1, P_1P_2, P_2P_2, P_1p, P_2p, pp
Lutheran	Lu^a, Lu^b	anti-Lu^a anti-Lu^b	Lu (a+b−), Lu (a−b+)	Lu^aLu^a, Lu^aLu^b, Lu^bLu^b
Kell	K (Kell) k (Cellano)	anti-K anti-k anti-Kp^a anti-Kp^b	K+K−, K+k+, K−k+, [K−k−]	KK, Kk, kk
Duffy	Fy^a, Fy^b	anti-Fy^a anti-Fy^b	Fy (a+b−), Fy (a+b+), Fy (a−b+), Fy (a−b−)	Fy^aFy^a, Fy^aFy^b, Fy^bFy^b, $FyFy$
Kidd	Jk^a, Jk^b	anti-Jk^a anti-Jk^b	Jk (a+b−), Jk (a+b+), Jk (a−b+), [Jk (a−b−)]	Jk^aJk^a, Jk^aJk^b, Jk^bJk^b
Diego	Di^a	anti-Di^a	Di (a+), Di (a−)	Di^aDi^a, Di^aDi, $DiDi$
Sutter	Js^a	anti-Js^a	Js (a+), Js (a−)	Js^aJs^a, Js^aJs, $JsJs$
Auberger	Au^a	anti-Au^a	Au (a+), Au (a−)	Au^aAu^a, Au^aAu, $AuAu$
Xg	Xg^a	anti-Xg^a	Xg (a+), Xg (a−)	Xg^aY, XgY, Xg^aXg^a, Xg^aXg, $XgXg$

We shall now review the knowledge derived to date from comparisons of some of these systems in man with those in non-human primates.

3. THE ABO AND THE LEWIS SYSTEMS IN MAN AND THE OTHER PRIMATES

The frequency of the phenotypes ABO vary in different populations. This fact has been of great interest to anthropologists, who immediately tried to use it to classify the various human populations. In less than fifty years abundant data have been amassed on this system and practically every human population has been studied, although not all of them in great detail.

The variations in the geographical distribution of the frequency of a given gene can be represented diagrammatically by drawing a line between the points of equal genic frequency (isogenic line) (Fig. 8.1).

Another method of representing the geographical variations for this character is that of indicating each population by a point in a system of

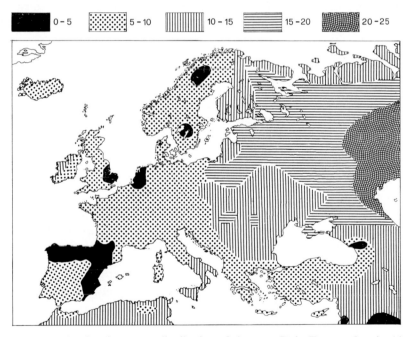

FIG. 8.1 Map showing frequency distribution of the gene I^B in Europe. One is able to detect a decrease in the percentage of gene frequency going from east to west. This could be a result of the frequent invasions of the Mongols; who are characterized by having a high frequency of this gene.

triangular co-ordinates on the sides of which are reported the frequencies of each of the three genes I^A, I^B, I^O.

A large majority of human populations has a frequency for the gene I^O which is greater than 50%. Any calculation of the average genic frequencies for the ABO blood groups for all of mankind must be open to question, but nevertheless such an estimate could be useful for seeing where the various populations fall by comparison and possibly for comparing the human frequencies with those of the non-human primates.

According to the most reliable calculations, the frequency of the gene I^A is 21·5%, that of the gene I^B 16·2% and that of the gene I^O 63·3%.

The highest frequency of the gene I^O is found among the American Indians, while the Eskimos have a relatively low frequency for the genes I^A and I^B as well (Fig. 8.2).

The frequencies of the genes for blood groups ABO in western Europe in general are 26% for gene I^A, 6% for gene I^B and 68% for gene I^O. There are, moreover, many regional variations of particular interest. In Great Britain, for example, the proportion of group A to group O is higher on the west coast (northern Wales, Scotland) than in the south and on the east coast. Rather high frequencies of O are also found in Ireland and in the populations of the eastern part of Iceland, differing in this respect from Norway, the country from which, according to history, the Viking colonizers of these regions came.

Another region in Europe where the frequencies of blood group O are particularly high, are the areas in Spain and south-western France where Basque is spoken. Basque is a language of obscure origin, certainly not belonging to the Indo-European group of languages. In Sardinia, too, there are particularly high frequencies of blood group O. The origin of the Sardinians, like that of the Basques, is very controversial; the study of the blood groups and of many other hereditary characters of these populations is of special interest as a means of discovering their possible origins.

The variant A_2 of group A is very frequent in Europe (about 10%) and in Africa, but is absent or very rare in other regions. The Lapps, who differ from the other inhabitants of Europe in many characters, also display a very high frequency of the gene A_2 (about 35%).

The Lewis system, discovered in 1946, is in some ways closely correlated with the ABO system. It is based on the secretion or the non-secretion in the saliva of the hydrosoluble substances of the antigens A, B and H (H denotes a group-specific substance present on the red corpuscles of group O secretor-individuals, which is demonstrated by means of the extract of the seeds of certain plants).

The frequencies of the genes for blood group ABO in human populations is not however our main topic of interest. Our main aim is to present the

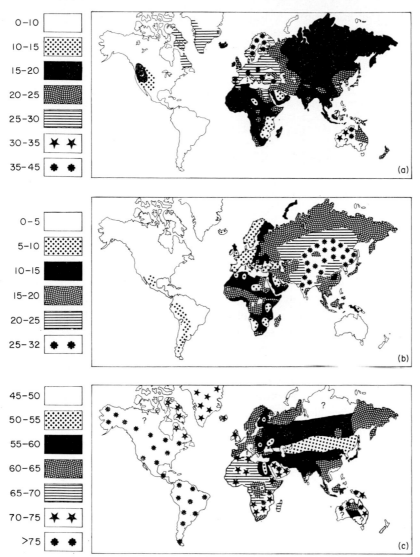

FIG. 8.2 Distribution of frequency of the gene for the blood groups A B O in Man: (a) the frequencies of blood groups O; (b) the frequencies of the gene I^A and (c) the frequencies of the gene I^B.

results of a comparative investigation in the different primate species.

Among the anthropoid apes the chimpanzee gives isoagglutination reactions indistinguishable from those of human blood. No serological differences have in fact been demonstrated by reciprocal absorption between chimpanzee and human group A red cells. Saliva derived from a random sample of 100 chimpanzees proved that they are secretors. As in man there are population differences in the distribution of the blood groups (see Table 8.3A).

TABLE 8.3A. Distribution of the human-type A-B-O blood groups among 4 sub-species of chimpanzees (from Wiener and Moor-Jankowski, 1971).

Subspecies of *Pan troglodytes*	O		A		Totals	Gene frequencies	
	Number	%	Number	%		O	A
P. troglodytes	2	14·3	12	85·7	14	37·9	62·1
P. schweinfurthi	13	39·3	20	60·6	33	62·7	37·3
P. verus	4	9·5	38	90·5	42	30·8	69·2
P. koola-komba	1	—	1	—	2	—	—
All chimpanzees tested, including those above	33	14·5	195	85·5	228	38·0	62·0

Isoagglutination patterns in orangutan and gibbon blood revealed that both have serological reactions indistinguishable from ABO groups of man. Among 26 orangutans 22 were of group A, 1 of group B and 3 of group AB. Their red cells however failed to react with anti-H reagents even though strong inhibition reactions for H were given by the saliva. Among 52 gibbons tested, Weiner and his coworkers did not find any with blood group O (there were 10 group A, 20 group B and 22 group AB). Estimated gene frequencies of the A allele is 0·40, and 0·60 for the B allele. All the gibbons from whom saliva was obtained proved to be secretors.

In contrast with the previous tests the gorilla failed to show any iso-agglutination reactions. Gorillas do not have A-B-H blood group substances on the surface of their red cells; they have their ABH substance inside the red cells and in the serum. All the 11 lowland gorillas tested by Wiener and co-workers had anti-A but not anti-B in their sera and therefore are of blood group B. Two mountain gorillas whose urine samples were tested by Candela were of group A.

As in the case of gorillas the Old World Monkeys fail to demonstrate clear blood groups on their red cells. The antibodies for the ABO substance exist in the serum and blood group substances in their secretions. Among the catarrhines some species such as baboons and crab-eating macaques exibit all 4 ABO phenotypes, other species, notably rhesus monkeys have only blood group B, while in others, such as patas monkeys, only group A was found. The distribution of ABO blood groups for 4 species of baboons is presented in Table 8.3b.

In the New World Monkeys and in the Prosimians tested, the principles derived for the Old World Primates seemed to apply with less regularity and their serological reaction for the ABH blood group indicated even less similarity with man.

TABLE 8.3B. A-B-O blood groups of baboons (from Wiener and Moor Jankowski, 1971).

Species of *Papio*	Blood groups					Estimated gene frequencies		
	O	A	B	AB	Totals	O	A	B
P. cynocephalus	0	18	20	22	60	8·9	44·0	47·1
P. anubis	0	53	56	65	174	8·4	45·0	46·6
P. ursinus	0	4	59	26	89	1·8	18·8	79·3
P. papio	2	27	93	66	188	10·6	29·9	59·5

Recapitulating, we may thus affirm that the substances of the ABO system are regularly present in the organs and in the secretions of all the primates tested to date, while the antigens are not always to be found on the erythrocytes. Some investigators have wished to see in this particular distribution (schematized in Table 8.4) a phylogenetic significance or at least

TABLE 8.4. Antigens present on the erythrocytes of some primate species.

Species	Antigens found on Erythrocytes		
Homo	A	B	H
Pan	A	—	H
Pongo	A	B	—
Gorilla	A	B	—
Hylobates	A	B	—
Symphalangus	A	B	—
Cercopithecidae	—	—	—
Cebidae	—	—	—
Lemuridae	?	?	?

a distinctive character of these groups of primates. Recent research has, however, modified this picture: in a number of individuals of apes, monkeys and prosimians, antigens of type A or B located on the erythrocytes and antigens of type H have been detected, which can be looked upon as a major discovery, having previously been known only in the chimpanzee and in man. Evidence of these antigens has also been brought to light in animals lower on the scale of evolution (even in bacteria).

The problem of the uniformity of the mechanism of genetic determination of the ABO system in the primates rests at this point. The problem cannot yet be attacked because of the scarcity of genetic data for the non-human primates, especially the less advanced monkeys. In the chimpanzee it is assumed that the mechanism is linked to a dominant allele for A and a recessive for O, analogous to the situation in man, while for the gorilla and the orangutan the presence of two isovalent genes, one for A and one for B, is postulated. An added factor is that from the data so far obtained it would appear that these two species are not in genetic equilibrium. This fact could be due to a non-random type of mating which favours the homozygotes, or to a particular geographic distribution of the various genotypes, or it could depend on other mechanisms which are not at present understood.

4. The MNS System in Man and the Other Primates

The first antigens of this system which were discovered were the M and N, which can be demonstrated using antisera obtained by immunizing rabbits with red blood cells. The inheritance of these antigens is very simple because it depends on a single pair of alleles, M and N, which can give rise to the three phenotypes M, MN and N. The genic frequency of a population can be estimated very easily by counting the genes directly.

Generally speaking, in Europe, the frequencies for the genes M and N are almost equal, and consequently approximately 50% of the people are MN heterozygotes. Nevertheless among the Sardinians the frequency of the gene M is remarkably high, reaching as much as 75%. Frequencies that are different from those generally found in Europe for this gene are also a distinctive characteristic of the Lapps.

In India and in southeast Asia the gene M is slightly more frequent than in Europe, but it decreases rapidly east of Java and Borneo as far as Australia where its frequency is less than 30%.

The highest frequency of this gene is found among the American Aborigines with values lying between 75 and 95%. Again in this case a marked contrast is seen between the American populations and those which at present live in eastern Asia.

To the original MN system the S factor has been added. The S gene is found associated either with M or N, so that a person may be MS, MNS or NS. It was discovered in 1947 in England where it is very common. In Europe the genic frequency for S is lower. Among the Chinese and the Malays it is only 3%. Although the frequencies of MN individuals in New Guinea and Australia are almost the same, the S gene is rare or absent in Australia, while among the Papuans of New Guinea it reaches 12%. Some tribes of American Indians have a frequency for S higher than 30%.

All the chimpanzees studied to date have an M-like antigen on their red corpuscles; about a third of them have the factor N^v. This factor (N^v) is one of the numerous N factors present in man, and can be demonstrated by using lecithin from *Vicia graminea* (Table 8.5).

About half of the 24 orangs studied exhibited an M factor comparable to the human one, so that orangs can be divided into types M^{or} and m^{or} respectively.

Among 24 gibbons examined for this system 14 were recognized as N, 5 as MN and 5 as M, and the serological characteristics of the M and N factors of the gibbons are considered to be more similar to those of human beings than are those of the other primates. Their genetic determinism also appears to be more similar than that of the chimpanzee or the orang.

TABLE 8.5. Frequency of M and N^v factors in some primates.

	M-like factors			N^v factors (*Vicia graminea*)		
	Present	Absent	Total	Present	Absent	Total
Chimpanzee	130	—	130	42	62	104
Gorilla	1	—	1	1	—	1
Orangutan	12	12	24	—	24	24
Gibbon	10	14	24	19	5	24

The regular presence of an M-like factor has been confirmed in the blood of the Old World monkeys, while in the monkeys of the New World only some of them show an M-like factor. N-like factors have never been observed either in the Old World or in the New World monkeys.

Recent experience seems to demonstrate that the human M and N antigens are formed of a mosaic of which only some components are present in the blood of the chimpanzee and the other primates and that this resemblance becomes gradually less as the zoological scale is descended. Thus it would seem that this factor is of great phylogenetic importance, but the problem is far from being clear.

5. THE RH SYSTEM IN MAN AND THE OTHER PRIMATES

In 1940 Landsteiner and Weiner discovered that the antisera produced by injecting a rabbit with the red blood cells of *Macaca rhesus* would agglutinate the blood of a large number of people. The Rh positive reaction was inherited as a dominant, the negative as a recessive. However, the genic interpretation of this system of blood groups is more complex than that as shown in Table 8.6.

The existence of the Rh factor explained, among other matters, the cause of the so-called "haemolytic disease of the newborn".

In almost all of Europe the frequency of Rh negative varies around 16%, corresponding to a genic frequency of 40% for the complex of recessive genes *c d e* (*r*). There are, nevertheless, exceptions, among which are the Basques who display a genic frequency of *c d e* falling between 50% and 60%, which is equivalent to an Rh negative phenotypic frequency of 30%.

Data for Africa are still not very numerous, but already it is possible to affirm that the frequencies of *c d e* are much higher than in any other part of the world. The highest frequencies have been found among the Nilotics and the Bushmen; populations which are very different physically from one another.

TABLE 8.6. Reactions found in the Rh blood group system following the Fisher-Race notation. a. Gene complexes frequently present in European populations. b. This antibody has not yet been identified.

Antibodies	Gene complexes[a]							
	CDE (R_2)	*CDe* (R_1)	*cDE* (R_2)	*cde* (r)	*cDe* (R_0)	*cdE* (R'')	*Cde* (R')	*CdE* (R_y)
Anti-C	$+$	$+$	$-$	$-$	$-$	$-$	$+$	$+$
Anti-D	$+$	$+$	$+$	$-$	$+$	$-$	$-$	$-$
Anti-E	$+$	$-$	$+$	$-$	$-$	$+$	$-$	$+$
Anti-c	$-$	$-$	$+$	$+$	$+$	$+$	$-$	$-$
(Anti-d)[b]	$(-)$	$(-)$	$(-)$	$(+)$	$(-)$	$(+)$	$(+)$	$(+)$
Anti-e	$-$	$+$	$-$	$+$	$+$	$-$	$+$	$-$

Among the Chinese and the Siamese the frequency of the genes *c d e*, and, therefore, of Rh negative, is rare. It is rarest of all among the Australian aborigines (Fig. 8.3).

The complex of genes for the Rh factor is especially interesting for comparative studies because it represents a "supergene" or a complex of closely associated genes which is present in many species. There are two alternative hypotheses, not necessarily mutually exclusive, concerning the way in which this complex of genes may have originated. The first is based on Fisher's hypothesis according to which the linkage between genes which alters their selective value reciprocally tends to increase. In conformity with this theory, the *C* and *D* of the Rh factor, which in man probably does alter their selective value reciprocally, might have originated on different chromosomes, and in succeeding stages would have become closely associated by means of translocations. The other possibility is that this complex of genes arose by means of duplication of genes followed by divergence of function. The Rh system in the anthropoid apes, and as far as is known in the other primates as well, exhibits some interesting differences with respect to that found in

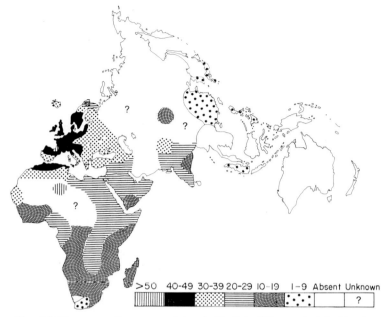

FIG. 8.3 Percentage frequency of the *r(cde)* allele (Rh negative) in blood.

our species, which at a later date may enable us to decide between these two hypotheses.

Rh+ and Rh— individuals are found among gorillas with the distribution of the subgroups resembling that among human beings (that is, all three subgroups are present). In the orang, on the other hand, the number of reactions is so low as to indicate that the factors for Rh in the orang must be very different from those of humans. In the gibbons there is only a weak capacity for absorption of anti-e, while no antigens of the Rh system exist in Old World monkeys, nor in those of the New World, nor in prosimians. It must be kept in mind, however, that these data have been derived from a very small number of individuals and, therefore, have a purely indicative value.

According to Franks (1963) the data for the anthropoid apes, together with the rare instances of human beings in which the product of the gene for the *E* locus is missing, could furnish an indication in favour of each of the series. One could also speculate that the *D*-like antigen represents the primordial Rh-antigen from which the others originated, because the *D*-antigen is the most diffused of the Rh-antigens among the primates.

Comparative research on the complex of immuno-reactions in the anthropoid apes will undoubtedly supply much useful information on how the antigen-producing genes and their modifiers are constituted but many lacunae in our knowledge of human blood groups could also be filled in by increasing our knowledge of the isoantigen systems of individuals of the primate species.

6. BLOOD GROUPS AND NATURAL SELECTION

For a long time it was held that the blood groups were not subject to selection; that is, that they were not adaptive characters, and on that assumption many anthropologists have attempted to classify and establish relationships between the diverse human populations. However, the very geographical distribution, with its gradients of frequencies of the alleles for the ABO blood groups in the various human populations, suggests the hypothesis that there may be some relationship with the environment. The Rh blood group and its connection with haemolytic disease of the newborn clearly demonstrates the effects of the selection. Every time that an Rh+ foetus is killed by the antibodies produced by its Rh- mother, there is a loss of D and d genes. In this way, with the succession of generations, these genes should become steadily less frequent in the population, to the point of vanishing or of diminishing to a frequency such as to be maintained only by the rhythm of mutation. Nevertheless, these frequencies are very high, especially in some populations. It is evident that some selective advantage is playing a role in maintaining this polymorphism.

The same thing may be said for the ABO blood groups. Recent research has demonstrated that there is a considerable death rate among foetuses of certain genotypes, especially where the mother is O and the foetus is A.

In recent years it has been discovered that the frequency with which individuals of a particular blood group are affected by certain diseases is different from that of the general population. The results are most striking in cases of gastric cancer and duodenal ulcers. People with group O blood, for example, are more subject to duodenal ulcers than others.

Satisfactory explanations have still not been found for these possible selective factors, but they certainly exist and have played diverse roles in the different conditions in which man has evolved throughout the ages. The comparison of these data with those of the other primates could perhaps help to interpret the polymorphism of the blood groups of our species.

7. SENSITIVITY TO PTC (PHENYLTHIOCARBAMIDE) IN MAN AND THE OTHER PRIMATES

Some people are unable to taste phenylthiocarbamide, a synthetic compound of phenylthiourea which among its organolectic characteristics has the quality of being as bitter as quinine.

Sensitivity to this substance is a strictly hereditary character of a simple Mendelian type.

The frequency of the dominant phenotype in man is about 80%. The differences in the frequencies of the genotypes among the various populations are nevertheless remarkable (see Chapter 3, Section 7).

For example, the frequencies of the non-taster phenotype in northwest Europe range from 35 to 40% and become much lower in the circum-Mediterranean regions. Among the Chinese, the Africans, the Japanese, the American Indians and the Lapps the non-taster phenotype is much less frequent.

The inheritance of the sensitivity to this substance has also been demonstrated among other primates. It has been shown, too, that man and the chimpanzee have an almost similar threshold of sensitivity. Population data gathered at the level of individual genera have revealed different frequencies of the taster phenotype in the various groups.

Within the ambit of the platyrrhine monkeys, for example, in the genera *Ateles and Lagothrix* there is a high frequency of non-tasters, while the opposite is found in the genera *Callithrix* and *Leontideus*, the frequency of the tasters in the genus *Cebus* is extremely low whereas it is very high for the genus *Saimiri*.

In the catarrhines the differences observed between the genera *Macaca*, *Papio*, *Theropithecus*, *Cercocebus* and *Cercophithecus* are minimal.

The frequency with which the taster characteristic is present in the genera *Colobus* is very similar to that of the other genera of African monkeys. However, a special situation is found among the anthropoid apes and gibbons. The genus *Hylobates* exhibits a frequency of tasters of 50%. The orangutan has an extremely low frequency (5%). *Pan* and *Gorilla* show a frequency of tasters more or less similar to that found in man and in the African and Indian monkeys (see Fig. 3.8).

These data, and particularly those relating to the platyrrhines and orangutan, indicate the possibility that mechanisms of genetic drift or of selection have intervened to lead to these all-or-nothing conditions. To explain these data, reference must be made to the chemical composition of this substance.

Many other chemical compounds similar to PTC give the same bimodal reaction, although PTC gives much clearer results. The part of the formula found to be essential in all of these substances which are active from the standpoint of sensitivity is:

$$C = S \begin{cases} NH - C_6H_5 \\ NH_2 \end{cases}$$

All the derivatives in which oxygen replaces sulphur are inactive.

The fact that certain thioureas, especially thiouracil, are used in hospitals to depress thyroid activity, suggests that the polymorphism of the gene for sensitivity may in some way be connected with thyroid function. Confirmation of this may be that the diverse frequencies for tasters and non-tasters are found to coincide in the same population with the effects caused by some dysfunctions of the thyroid.

Further, the fact that chemical substances similar to PTC are normally found in some vegetables (Brassicacae), and the presence of this type of polymorphism in the other primates supports the idea that this condition has been transmitted for a long time in the process of evolution and has been maintained by means of a heterozygotic mechanism, possibly originating in dietary factors which influence the thyroid gland. The altered activity of the gland would have led to the selection of one allele or the other; or would have maintained both in equilibrium if selection did not eliminate one.

GENERAL REFERENCES

Buettner-Janusch, J. (ed.) (1963). "Evolutionary and Genetic Biology of Primates". Vols 1–2. Academic Press, New York and London.

Buettner-Janusch, J. (1966). "Origins of Man. Physical Anthropology". J. Wiley & Sons, New York, London and Sydney.

Chiarelli, B. and Scannerini, S. (1963). I Gruppi sanguigni ABO degli antropoidi attraverso una comparazione con quelli dell'uomo. *Arch. Antrop. Etnol.* **93**, 1–13.

Chiarelli, B. (1963). Sensitivity to P.T.C. (Phenyl-Thio-Carbamide) in Primates. *Folia Primatologica* **1**, 103–107.

Franks, D. (1963). The blood groups of Primates. *In* "The Primates". Symposia of the Zoological Society of London, no. 10, Academic Press, London and New York.

Kalmus, H. (1971). Phenylthiourea Testing in Primates. *In* "Comparative Genetics in Monkeys, Apes and Man" (B. Chiarelli, ed.), Academic Press, London and New York.

Lush, I. E. (1966). "The Biochemical Genetics of Vertebrates except Man". North-Holland Publishing Co., Amsterdam.

Neel, J. V. and Schull, W. J. (1954). "Human Heredity". University of Chicago Press, Chicago.

Race, R. R. and Sanger, R. (1962). "Blood Groups in Man". Blackwell Scientific Publications, Oxford.

Wiener, A. S. and Moor-Jankowski, J. (1971). Blood Groups of Non-human Primates and Their Relationship to the Blood Groups of Man. *In* "Comparative Genetics in Monkeys, Apes and Man" (B. Chiarelli, ed.), Academic Press, London and New York.

9

A Comparison between Serum Proteins and Immunological Distances

1. GENERAL

Recent experimental and theoretical developments in the chemistry of proteins and in genetics have made it possible to establish that the specific structure of natural proteins is subject to rigorous genetic control. Each protein has a surface configuration made up of various active centres which are responsible for its functional activity (whether as a hormone, an enzyme or an antibody). It may be assumed that each of them is genetically determined and is specific for each individual and for each species or group of species. In other words, we may say that the genetic information expressed in a single protein reflects the evolutionary history of the organism. We have already seen this in connection with haemoglobin.

For the foregoing reasons a comparative study of serum proteins should furnish valuable information on the phylogenetic relationships of any determinate group of organisms. What is more, it is sometimes possible to derive information regarding the speed of evolution of the different parts of the genic complex of these organisms; that is, it can be deduced whether or not certain proteins and their corresponding genes have evolved more rapidly than others.

Moreover, since many proteins or complexes of proteins are formed by the combined action of several genes, the study of proteins with the object of reconstructing the evolution of a species will yield more information than the study of the frequencies of single genes.

2. THE TYPES OF PROTEINS IN HUMAN SERUM
AND THEIR VARIANTS COMPARED WITH THOSE OF THE
NON-HUMAN PRIMATES

The corpusculated bodies of the blood are suspended in a liquid of different specific gravity (1090 for the corpusculated elements as against 1030 for the liquid plasma) and can, therefore, be easily separated. In this way the liquid plasma, a faintly yellow fluid, can be obtained.

Plasma is composed of 91–92% water and 8–9% of other substances, of which approximately 8·5% are proteins and the others are lipids, glucose, steroids and inorganic compounds.

The serum is separated from the plasma by coagulation. It differs from plasma only in that it lacks fibrinogen. The plasma proteins thus fall into two groups, the one comprising fibrinogen and the other the true serum proteins (serum albumin and serum globulin) which do not lose their original solubility when coagulation takes place.

Fibrinogen constitutes about 4% of the total plasma proteins. The remaining serum proteins (81%) in our species are distributed as follows:

serum albumin	64–70%
serum globulin α_1	3–4·5%
serum globulin α_2	6–9%
serum globulin β	8–12%
serum globulin γ	9–13%

There are various techniques and methods used to demonstrate the kinds of proteins found in blood serum. One of the most efficient and popular is that of electrophoresis. In an electrophoretic field all the proteins of the plasma migrate to the anode at different speeds. When a drop of serum is placed on a piece of filter paper and placed in an electrophoretic field, and then is stained, bands representing the various serum proteins begin to appear.

The bands are due, consecutively, to albumin, α–globulins, β–globulins, fibrinogen and γ–globulin. In such an electrophoretic diagram however, not more than six bands are clearly evident; in fact, there are many more types of protein actually present in the plasma. The fractions of the α and β globulin are really made up of groups of proteins which possess approximately the same electrophoretic mobility if identical experimental conditions are adopted. A better electrophoretic separation can be obtained using agar gel (Fig. 9.1).

The albumins exhibit an anodic mobility $(^+)$ and are preceded by a more rapid component known as prealbumin; all the other components migrate towards the cathode $(^-)$.

According to Smithies, vertical electrophoresis using starch gel results in more discriminating separation of the different components and about 20–30 may be perceived using this method.

Actually, because each protein requires different experimental conditions (for instance different buffer pH, gel density, temperature, electric intensity, etc.) it is fruitful to study only one protein at a time. Some of the molecular

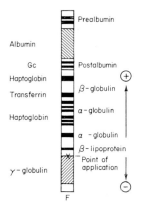

FIG. 9.1 The relative electrophoretic mobility of proteins found in normal human blood serum. It has been found that different electrophoretic conditions are required to obtain the best results with each protein. However, experimental conditions often used are: a voltage of 7–20 V/cm; temperature from 5°–20°C.; timing from 10 to 24 hours; pH from 8 to 9; and a saline concentration less than 0·05 M. In order to obtain reproducible results, the same experimental conditions must be rigorously kept for each sample of serum, and for each electrophoresis.

groups present in blood serum exhibit hereditary individual variations which represent very interesting kinds of polymorphisms.

We shall now examine separately the different types of molecules found in the serum of man and in non-human primates.

1. *Albumin.* Albumin is the only simple protein which is found in the serum. Its molecular weight is low: about 67,000. It is responsible for the maintenance of the osmotic pressure of the blood (25–30 mm Hg); moreover it acts as a protein reserve for the organism and as a carrier of substances which are insoluble in water. These substances become soluble immediately upon being bound to albumin.

Only recently hereditary variations that have a relatively more rapid electrophoretic mobility have been described in the serum albumin in our species (Naskapi Indians of Canada). As far as the other primates are concerned only a single variant is known. This is found in *Papio anubis*, in which a slower variant has been noted in 30% of the 54 animals studied.

2. *Group specific proteins* (Gc). These proteins appear as one or two bands behind the albumin band in acrilamide gel electrophoresis. The

function of these proteins is unknown, however, they are frequently used in the study of human genetics because they are polymorphic variants. Using immunoelectrophoretic techniques it is possible to distinguish 3 Gc phenotypes in man. A polymorphism identical to that of man has been found in the orangutan, but not in chimpanzees. Some baboons also seem to show a similar polymorphism for this type of serum protein.

3. *Lipoproteins.* α–globulins are lipoproteins in which the prosthetic groups make up about 75% of the molecule. They seem to act as vehicles for the transport of compounds insoluble in water, and in fact much of the cholesterol in blood is tied to the lipoproteins.

The α–lipoproteins are more plentiful in the plasma, but have a much lower molecular weight (about 1/6 as great). Hereditary variants have been described for human α lipoproteins and antibodies produced for these in rabbits have resulted in the identification of an antigen: Lp (α). The same kind of human polymorphism with positive and negative reactions to the human anti-serum anti Lp (α) has been described in various chimpanzees, orangutans and macaques which have been tested. Another hereditary variant found in man (Ld), has not been found in non-human primates.

4. *Transferrins.* The most abundant β globulin in the serum is the glycoprotein β_1–transferrin (siderophyllin). It is capable of binding 1·25mg of Fe^{3+} per mg (that is, two atoms of metal per molecule) and in certain conditions (lack of iron and pregnancy) its concentration can increase remarkably. The prosthetic group represents about 5·5% of the whole molecule. Recent research has demonstrated that the β_1–globulin fraction of the serum is made up of three components, each with a different anodic mobility which, were recognized as transferrins by means of autoradiographic techniques using $^{59}FeCl_3$ (Fig. 9.2). The three types of transferrins are probably allomerous forms of a single protein chain.

The most common genotypes are governed by a series of apparently allelic genes; genealogical studies indicate that they belong to the same locus.

At the moment more than 20 polymorphic variants of the transferrins are known in man; some of which are extremely rare. However their functional capacities are more or less equal, both in regard to the transportation of Fe^2+ and also their activity in defence against infection.

Other species of animals (chickens, mice, pigs, most of the primates) exhibit an even greater degree of polymorphism than that found in the human species.

Polymorphism in the transferrins has been found in the gorilla and in the chimpanzee. In a group of 111 chimpanzees Goodman (1967) was able to distinguish five different types of transferrins. These can be related to the same number of alleles.

Many different transferrins have also been found in some species of macaques: approximately ten in *Macaca mulatta*, eight in *Macaca irus*, two in *Macaca fascicularis*. In contrast, fewer variations seem to be present in baboons.

A general comparison of the transferrins of the primates indicates that those of the Hominoidea and of the Ceboidea have, as a group, less electrophoretic mobility than those of the Old World Monkeys.

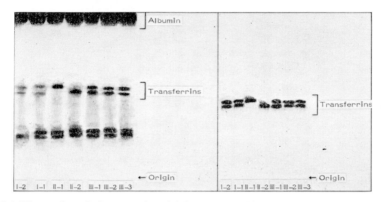

FIG. 9.2 Electrophoresis in a starch gel (diagram on left), and autoradiography through ^{59}FeCl$_3$ (diagram on right), showing some variants of transferrin (redrawn from a diagram in Schultz and Heremans, 1966).

5. *Haptoglobins* (Hp) α_2–globulin, a glycoprotein, is the protein found in highest concentration in human plasma (about 100 mg in every 100 ml). Its molecular weight is about 85,000, of which the prosthetic group represents approximately 23%. The haptoglobins are capable of binding haemoglobin when it is injected or freed by haemolysis, preventing it from damaging the renal tubules.

By using starch gel electrophoresis Smithies has identified three types of haptoglobin: Hp 1–1, Hp 2–1 and Hp 2–2 (Fig. 9.3). In persons who are homozygous for the gene Hp only a single band of components which bind haemoglobin is seen in electrophoresis. The genotype of homozygotes Hp^2 (Hp 2–2) displays a characteristic electrophoretic profile made up of a series of bands of decreasing mobility and concentration. These seem to represent polymers of a single polypeptide chain. The genotype Hp 2–1 heterozygote exhibits a mixture of heterogeneous molecules, of which only a small part behaves exactly like Hp 1–1 when subjected to ultracentrifugation. The remaining kinds of molecules form a series of polymers which have a mobility differing from that of the homozygote.

Actually Hp is formed of 2 different types of sub-units, which are called chains α and β. The β chains are the same in the three types of haptoglobin described above, while the α chains are different in all three; in Hp 2–1 the two types found in the homozygotes appear in the same proportion.

Moreover, the gene Hp1 has two allelic forms, Hp1F and Hp1S. Individuals homozygous for Hp1F have chains of a clearly higher mobility than those of Hp1S homozygotes, while Hp1S/Hp1F heterozygotes possess both types of chain in the same proportion.

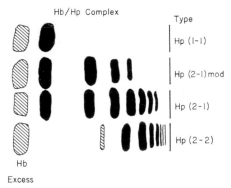

FIG. 9.3 Haptoglobin variants found in normal human serum (the reaction of peroxide with benzidiene takes place on the haptogolbin/haemoglobin complex after electrophoresis in a starch gel).

The genic product α 1F (the polypeptide chain synthesized under the control of the gene Hp1F) and that of α 1S have identical molecular weights, and the same –N– and –C– terminal amino-acids, but differ in one peptide: the central one being respectively N, F, C and N, S, C. The α 2 has both F and S and furthermore has one not found in α 1, the peptide J. The latter is composed of pieces of N and C fused together, even though these'two peptides are also present (Fig. 9.4).

It is possible that the gene Hp2 has evolved from a fusion of the primitive genes Hp1F and Hp1S. According to Smithies' hypotheses, fusion of this type could have occurred in heterozygote individuals by non-homologous crossing-over during meiosis (Fig. 9.5).

If this is true the gene Hp2 must be the result of a fairly recent mutation. In fact all the mammals examined to date, including the anthropoid apes, possess only the monomer form of haptoglobin similar to human Hp 1–1.

The genic frequencies of the different types of Hp also have a precise geographical distribution. The Hp1 alleles increase in frequency from east to west starting in Kashmir where their frequency is minimal.

On the basis of these findings, some authors have thought that this region was the centre of origin of the gene Hp2. However, Smithies' system makes it more plausible that in this area the first genic fusion giving rise to Hp2

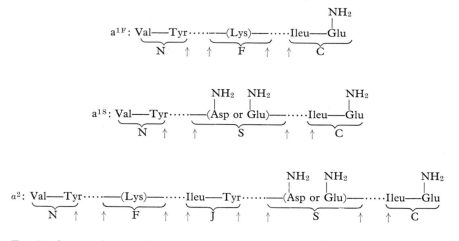

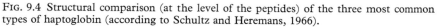

FIG. 9.4 Structural comparison (at the level of the peptides) of the three most common types of haptoglobin (according to Schultz and Heremans, 1966).

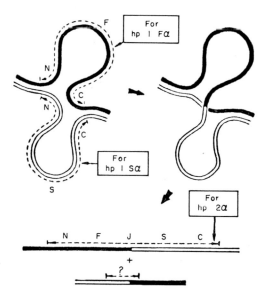

FIG. 9.5 Scheme for unequal crossing-over (following Smithies) in a heterozygote Hp^{1F}/Hp^{1S}; also, the linear form of the gene Hp^2. Note the region of the chromatid which supposedly directs the synthesis of the peptide.

took place, and that Hp2 is now in a phase of expansion and is replacing the Hp1 gene. Thus, according to this theory, the polymorphisms present in the haptoglobins of the human species must represent an unstable phase (transient) in which one of the homozygotes has a clear advantage.

Populations do exist in which the polymorphism of Hp is perfectly balanced: that is to say the relationship of the genic frequencies does not vary with the passing generations. In these cases it must be remembered that the heterozygous condition is probably in some way favoured by the environment, as in the instance of HbS in malarial zones. It is significant that haptoglobin plays an important role in porphyrin catabolism.

6. *γ–globulin.* The γ–globulins are glycoproteins in which the prosthetic group makes up not more than 3% of the entire molecule. From the chemico-physical point of view they are indistinguishable from the antibodies of the blood.

It is likely that γ–globulin and antibodies differ from each other only in the spatial disposition of the peptide chain. Nevertheless it is difficult to reconcile the enormous number of antibodies which an individual can produce with the few genes involved, at least in the germinal line to the synthesis of the γ–globulins. That is, it is a question of finding out how a relatively restricted number of structural genes can control the synthesis of so many proteins. The problem is still wide open. According to some authors this variability in the product could be brought about by means of the mechanisms of incomplete duplication and non-homologous crossing-over already seen in the haptoglobins. In other words the theory of these authors is that the different antibodies are synthesized in cells which have become diversified by means of autosomal rearrangements involving the structural genes of the γ–globulins.

These chromosomal rearrangements presumably would occur during mitosis leading to the formation of the mature immunization system. Since the pairing of the homologues and inter-chromosomal crossing-over are extremely rare events in mitosis, it might be postulated that these rearrangements take place between chromatids of the same chromosome. This is a well-known occurrence in some vegetables.

The immunoglobulins are all characterized by a common structural plan, consisting of two pairs of polypeptide chains: two light and two heavy (chains L and H). In this regard it is interesting that while there exist only two antigenic types of light chains, λ and κ, at least nine different types of heavy chain are known. Of the five kinds of immunoglobulin known (IgA, IgD, IgE, IgM, and IgG), undoubtedly the most important, quantitatively and qualitatively, is IgG. In the IgG fraction have been discovered two polymorphisms (Gm and Inv factors) each of which is controlled by alleles at independent loci. The control and synthesis of the heavy chains and light chains are regulated by the alleles of the Gm and Inv components respectively. The immunoglobulin which carries on important activity (rapid immunization) at the level of the mucous membrances is IgA. The most ancient immunoglobulin, present in all sera so far examined, is IgM, while the most recent (phylogenetically speaking) is IgD. The latter seems to be present only in man.

The data on the non-human primates concerning this genic complex are rather scarce and inconsistent. Nevertheless the heavy chain of the monkeys seems, in general, to be serologically very different from that found in human beings. Chimpanzees appear to be polymorphic for at least six of these factors, although the combination of them seems to differ from that found in human populations.

At least six Gm factors have been found in the gorilla, even though few animals have thus far been studied. The discrepancies in data obtained for the orangutan are too great to permit reference to them here, despite the fact that some similarities have been found in relation to human factors. It seems evident that the anthropoid apes, especially the chimpanzee and the gorilla, have a large number of Gm factors in common with man, and are certainly polymorphic for some of them.

The results obtained for the many species of Old World monkeys and some of the New World are difficult to synthesize briefly. Generally speaking the monkeys have considerably less specificity for human Gm than do the anthropoid apes. For instance baboons give negative results for Gm with six agglutinative sera, but a positive one for the seventh. This clarified a heterogeneity in the reagents which did not appear in studying the human material, and moreover demonstrates how useful experiments on different primate species are for the discovery of unusual characteristics in man.

3. Two-dimensional Electrophoresis of Serum in Primates

Two-dimensional electrophoresis, which combines separation on filter paper with separation in a starch gel perpendicular to this, permits resolution of a sample of mammal serum into 20–25 components. Thus this technique makes it possible to obtain electrophoretic profiles which invariably show a zone of prealbumin, a large albumin fraction, a group of 3–4 post–albumins, transferrins, haptoglobins, plus the S α_2 and γ globulins. This is illustrated in the diagram, Fig. 9.6.

The position of each fraction in two-dimensional starch gel electrophoresis is characteristic for each species studied. Individual variations are known only for the fractions which have a polymorphic determination, as transferrin or haptoglobin in man.

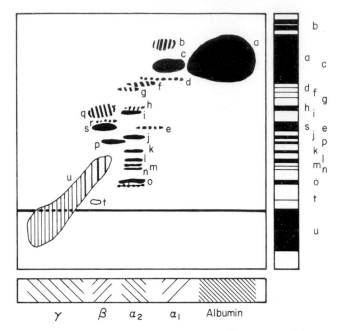

FIG. 9.6 Schematic representation of the electrophoretic pattern of human serum in horizontal, vertical, and two-dimensional electrophoresis in a starch gel.

The comparison of the diverse patterns can be used to establish extremely detailed phylogenetic distances and similarities at the taxonomic level. The two-dimensional electrophoretic patterns of the serum of the different species belonging to the same genus are generally very similar, if not identical. If we were to compare the electrophoretic patterns of individuals belonging to different genera, we would find greater variation.

Comparative studies of the sera of different species of the genus *Lemur* using this electrophoretic method have shown that three sub-species of the species *Lemur fulvus* have absolutely identical patterns, but these are clearly distinguishable from those of *Lemur variegatus* and *Lemur catta* in some characteristics. With the same technique it has also been possible to demonstrate the existence of completely different patterns in *Lemur*, *Propithecus* and *Galago*.

The electrophoretic picture of different species of the same genus (*Macaca mulatta*, *M. sylvana* and *M. irus*; *Papio comatus* and *P. hamadryas*) while not identical, still shows a general resemblance to one another. On the other hand, when comparing the electrophoretic patterns of individuals belonging to different genera, even of the same subfamily (*Hylobates* and *Symphalangus*), more marked differences are almost always found (Fig. 9.7).

However, in the genera *Macaca, Papio, Cercopithecus,* and *Theropithecus* whose electrophoretic profiles closely resemble each other is found an exception to this rule. In this regard two fractions are observed in the electrophoretic patterns of *Cercopithecus* and *Theropithecus* which are characterized by a mobility on paper corresponding to that of albumin and by an average mobility on starch gel.

The similarity in the two-dimensional electrophoretic profiles of this group of primates demonstrates that a high degree of genetic correspondence exists. This in turn suggests that these genera may have had a comparatively recent common ancestor.

The greatest differences from a qualitative point of view in this group of Cercopithecinae are those found in the α–globulin zone between the transferrins and the albumins in both paper and starch gel electrophoresis.

Presbytis entellus, the only species of the Colobinae which has so far been studied, presents an electrophoretic picture not very different from that of the Cercopithecinae.

Comparison of the two-dimensional electrophoretic profiles of the serum of the Hylobatinae with that of the Ponginae and the Cerocopithecinae reveals clear-cut characteristics which make it possible to distinguish the Hylobatinae from the others. Notwithstanding this, *Hylobates* and *Symphalangus*, as has already been noted, are quite different from each other, especially in the $S\alpha_2$—globulin zone.

The electrophoretic patterns of the true apes (gorilla, chimpanzee and orangutan) display differences between them that are even more marked than those seen between the genera of Cercopithecinae. Even though each anthropoid ape differs noticeably from the others, the gorilla and the chimpanzee have a group of at least ten fractions in the electrophoretic profiles of their serum proteins that are also common to man. These fractions are characterized by a high speed of migration which is greater than that seen in the orang.

These qualitative differences found in the two-dimensional electrophoretic patterns of the serum proteins of the various primates make it possible to advance ideas on the degree of genetic homogeneity among the various species. Results recently obtained from the study of the karyotypes of Old World primate species have amply confirmed these deductions. However, it is not possible to judge the degree of phylogenetic relationship between the different species from them. In fact so far we do not have indisputable principles which can be used to evaluate the frequency and direction of the mutations which may have resulted in the differentiation of primate serum proteins during the course of their evolution.

Nevertheless, these data can be utilized though with some reservation, in order to give a quantitative value to the phylogenetic and taxonomic

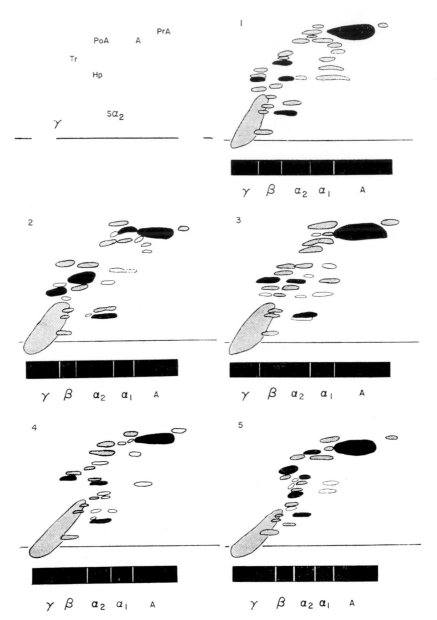

FIG. 9.7 Diagrams of two-dimensional electrophoresis in a starch gel of the serum proteins of Pongidae and some Cercopithecoidea: 1. *Gorilla gorilla*; 2. *Pan troglodytes*; 3. *Pongo pygmaeus*; 4. *Hylobates lar*; 5. *Symphalangus syndactylus*;

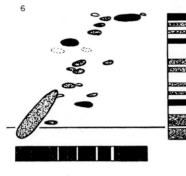

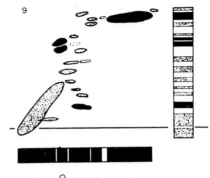

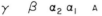

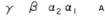

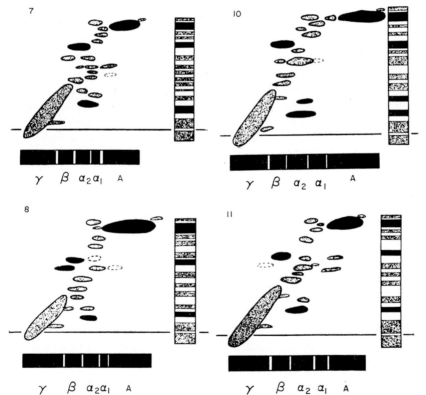

FIG. 9.7 (*continued*)   6. *Presbytis entellis*;
7. *Cercopithecus mitis*; 8. *Papio comatus*; 10. *Papio hamadryas*; 11. *Theropithecus gelada*.
In the upper left is a schematic representation for the identification of some of the fractions
(PrA, prealbumin; A. Albumin; PoA, postalbumin; Tr, transferrin; Hp, haptoglobin;
Sα_2, (slow) α_2—globulin; γ-globulin).

distances between the various species of primates. In fact, if we superimpose the electrophoretic pattern of a given species on a Cartesian pair, the axes of which are parallel to the directions of migration on paper and on gel respectively, we can measure the co-ordinates of the centre of each spot. Quantitative data of this nature can be arranged in matrices and elaborated following the method proposed by Sokal and Michener. Using these techniques of calculation it is possible to describe the taxonomic relationships existing between the various primate species entirely on the basis of the quantitative data on the electric charge of the serum proteins. Some of the results obtained to date agree well with the traditional classification, and underline still more natural groupings.

4. Immunological Distances

The immunological distance between man and the anthropoid apes (and other animals) was investigated by Mollison some time ago by means of so-called "serological reactions". When serum from human blood is injected into a rabbit, substances capable of precipitating the human serum proteins (precipitins) are formed in the serum of the rabbit. A serum thus prepared is called anti-human and is able to precipitate not only human serum protein (albumin was used by Mollison), but also that of other animals. In a similar way it is possible to obtain sera that are anti-chimpanzee, anti-macaca and so on.

It follows that each serum prepared in this way causes precipitation of maximum intensity with the serum of the species against which it is immunized (homologous reaction), and precipitation of lower intensity with blood sera of the other species (heterologous reaction). The first one who applied this type of immunological reaction to primate taxonomy, when the knowledge of protein structure was still quite rudimentary was Nuttall (1904). He found that the anti-human antisera produced total precipitation when tested with human serum, but the reaction was not as complete in chimpanzees and even less so in various monkeys and prosimians. Lately other researchers have used more refined techniques for this type of work.

Mollison (1912), in particular, noticed that an anti-human serum gives precipitates of a progressively decreasing intensity with the sera of the following species: man, chimpanzee, gibbon, orang, Cercopithecinae, Cebinae, prosimians and other mammals.

Experiments of this nature are generally carried out by mixing a constant quantity of antiserum with a progressively increasing quantity of serum. If the reaction is brought about in a graduated capillary tube of known volume it is easy to calculate the quantity of precipitate obtained. The intensity of the heterologous reactions is given as a percentage of the maximum intensity of the homologous reaction.

According to Mollison, the different intensities of precipitation are not so much due to quantitative as to qualitative differences between albumins contained in the sera of the blood of the diverse species. That is the "kinship reaction" becomes more intense as the two species being examined come closer to possessing an equivalent unity of albumin structure. This unit was called a "proteal" by the author. Serum immunized against that of one species will precipitate all the albumin contained in the blood of that species but in the serum of a different species only the albumin (the proteals) which this species has in common with the species used for the preparation of the antiserum will be precipitated.

Therefore the different intensities of precipitation seem to show that the serum albumin of man is closest in affinity to that of the chimpanzee and that the affinity then decreases throughout the animal series.

That man and the anthropoid apes possess common proteals (the term albumin is probably more appropriate) as well as others which are unique to each species has been demonstrated by testing the respective sera against the heterologous anti-sera. The results are commonly called "reciprocal reactions".

It has already been stated that serum X forms a precipitate of maximum intensity with anti-X serum and less with others. The reaction of human serum with anti-chimpanzee antiserum forms a total precipitate of almost 63%. The reciprocal reaction (serum *Pan*/antihuman antiserum) gives a precipitate of 85%. This result tells us that Man has about 2/3 of his proteins in common with *Pan* and 1/3 are strictly his own. *Pan* has 6/7 of its proteins in common with man and only 1/7 distinctly its own.

Table 9.1 indicates the intensity of precipitation (as a percentage) between sera and antisera of different species and the reciprocal reactions. In Fig. 9.8 are represented the data reported in Table 8.1. To understand the diagram it must be noted that the height of each column corresponds to the

TABLE 9.1. Intensity of precipitation in percentage between sera and anti-sera of diverse primate species, and the reciprocal reactions (data from Mollison).

Antiserum	Serum				
	Homo	*Pan*	*Pongo*	*Macaca*	*Papio*
Homo	100	85	71	64	65
Pan	63	100	85	74	72
Pongo	39	44	100	42	48
Macaca	45	46	34	100	83
Papio	40	42	34	61	100

quantity of the precipitate obtained in the homologous reaction (intensity 100%). The transverse lines in the lowest part of the columns indicate the largest amount of precipitate from heterologous reactions (expressed as a percentage of the homologous ones) and the upper portion outlined indicates the part of the protein that is species specific, i.e. that was formed after the separation of the species in question from the common ancestor.

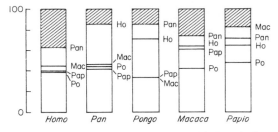

FIG. 9.8 Graphical representation of the data seen in Table 9.1: the intensity of precipitation (in percentage) between the sera and anti-sera of different primate species is shown.

Almost analogous results have been obtained more recently (1966) by Hafleigh and Williams. In Table 9.2 are synthesized some of the results on the degree of interaction of sera of various primates with antisera of human serum albumin produced in rabbits. The results were obtained by measuring the quantity of precipitate formed during the titration of the serum with the antiserum. Attempts were made to show the degree of similarity between the albumins of the various non-human primates and human albumins. The albumins of the chimpanzee and of the gorilla were seen to be very similar to human albumins, while those of the gibbon were less so, and the value for this species must be included within the limits of variability of the species of Old World monkeys tested. This, like other evidence that we shall consider in the following chapters, tends unequivocally to differentiate the Hylobatidae from the anthropoid apes. The values reported for the monkeys of the New World are lower, and those for the prosimians lower still.

5. DETERMINATION OF THE DEGREE OF ANTIGENIC CORRESPONDENCE BETWEEN HOMOLOGOUS PROTEINS IN THE SERA OF DIFFERENT SPECIES OF PRIMATES

Visual comparisons of this type can be integrated with an antigenic identification of the different protein fractions.

This method, which makes use of the antigenic properties of the homologous proteins in the sera of the diverse species of primates furnishes the clearest measure of the relative degree of correspondence of antibodies

TABLE 9.2. Immunological resemblance of the albumin of various primate species to human albumin. (From Hafleigh *et al.*, 1966.)

	Species	Similarity in %
Hominoidea	*Homo Sapiens*	100
	Pan	97
	Gorilla	92
	Hyloletes	79
Cercopithecoidea	*Macaca*	82
	Papio	79
	Cercocebus	79
	Erythrocebus	76
	Cercopithecus	79
	Colobus	80
Ceboidea	*Aotes*	74
	Ateles	58
	Saimiri	60
	Lagothrix	61
	Cebus	45
	Saguinus	54
Prosimii	*Lemur*	37
	Propithecus	22
	Potto	31
	Galago	28
	Tupaia	31
Other mammals	*Porcupine*	17
	Pig	8

among the various primates for each type of protein. This is especially evident if the analysis is carried out with many antisera on a particular type of protein.

If, for example, an animal fairly closely related to man, such as the macaque, is immunized with a human protein, only a small proportion of the superficial configurations of the proteins (the active centres) will produce an antigenic reaction, and these will tend to have a configuration which is strictly species-specific. If, instead, the animal being tested is phylogenetically very distant from man (such as a chicken) and is immunized with the same protein, all the superficial configurations of the protein will have antigenic power. The antibodies of some of these will be able to react with the antigenic configurations of other mammals outside the primate group.

This type of research has been recently developed particularly by Goodman. He has produced antisera from chickens, monkeys and rabbits for various human proteins and by utilizing special techniques has used them to prove their reactivity with the sera of many species of primates.

The experiments in immunodiffusion were carried out using Ouchterlony plates, which consist of a trefoil arrangement of three wells in which are placed the antigen preparations and the antiserum so that they diffuse into a central field of agar. Normally the antiserum is put in the bottom well and the two antigen preparations in the side wells (Fig. 9.9).

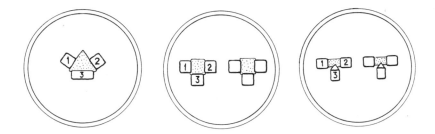

FIG. 9.9 Ouchterlony plates used for the study of immunodiffusion of serum proteins.

If one of the antigen preparations comes from the same species which furnished the antiserum (A_1) but on the other hand the second (A_2) is different, the precipitin line between A_1 and the antiserum will continue to grow even after it has met the precipitin line of A_2. This is due to the active centres of antigen A_2. This extension of the precipitin line is called a "spur", and the smaller the number of antigenic sites which A_2 has in common with A_1 (i.e. the greater the phylogenetic distance between the species furnishing the antiserum and the one being examined), the longer the spur will be (Fig. 9.10).

These comparisons clearly demonstrate the close phylogenetic relationship between man and the anthropoid apes; a lesser degree of relationship

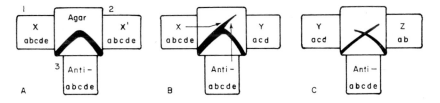

FIG. 9.10 Interpretation of results in experiments to show immunodiffusion in Ouchterlony plates. A, Identical reaction; B, Partial reaction with the formation of a spur on one side (the Y antigen has only the configuration acd in common with X; the antibodies for b and e are not precipitated from the Y molecule; they continue to diffuse through the agar until they meet the X molecule with which they precipitate to give the spur indicated by the arrow); C, Partial reaction with the formation of bilaterial spurs.

first with the Cercopithecidae, then with the Cebidae, and finally a much greater separation from the prosimians.

The results of analyses using the anthropoid apes, though incomplete seem to demonstrate that man is more closely related to the African anthropoids (gorilla and chimpanzee) than to those of Asia (orangutan). For example, in the case of chicken antiserum tested with human ceruloplasmin, and chicken and rabbit antisera tested with human transferrins, the gorilla and the chimpanzee appear to be identical to man, while the orang exhibits some differences. Testing macaque antisera with human albumin shows that gorilla and man are very similar, the chimpanzee differs only slightly from man and the orangutan displays marked differences from man and the other anthropoids.

Tests using chicken and rabbit antisera with human gamma-globulin show that in this case only the chimpanzee appears similar to man; the gorilla and orang are quite different from one another, man, and the chimpanzee.

Thus these data, as well as demonstrating that the African anthropoids (*Gorilla* and *Pan*) are more similar than the Asian one (*Pongo*) to man, also demonstrate that the phyletic separation of the orangutan is quite ancient.

Moreover, immunological data on *Hylobates* and *Symphalangus* seem to demonstrate a high degree of affinity in these species. According to a classification proposed by Goodman on the basis of his immunological data, *Homo*, *Pan* and *Gorilla* should be considered as belonging to the family of the Hominidae, while *Pongo* would be the only genus in the family Pongidae. Goodman considers *Hylobates* and *Symphalangus* very different from the anthropoid apes and puts them in the family of the Hylobatidae.

He has also found that among the Old World monkeys, the Colobinae are clearly distinguishable from Cercopithecinae. The results obtained also stress the difference between the Lorisoidea and Lemuroidea. The differences existing in this latter case however are not so great as those found between Tarsoidea and Tupaioidea.

The phylogenetic relationships of other groups of primates have also been analysed using immunochemical techniques. Pasika and Korinek, using anti-macaque sera obtained from rabbits, demonstrated that a higher degree of antigenic correspondence exists between the Hominoidea and the Cercopithecoidea and Ceboidea.

These immunological data show the validity of a taxonomic classification which proposes a closer phylogenetic relationship of the Ceboidea with the Cercopithecoidea than with the Lemuroidea and Lorisoidea, even if the separation of the protohominids probably happened at a very ancient stage in the evolution of the primates of the Palaeocene or early Eocene.

Other comparisons have been made using anti-potto and anti-lemur sera.

In every case considered the Lorisidae (*Galago, Loris, Potto*) and the Lemuridae (*Lemur*) show a greater antigenic correspondence among themselves than they do with the Tupaiidae or the catarrhine monkeys. This suggests that the phylogenetic separation between the Lorisiform prosimians (*Loris, Potto*) and the Madagascar form (*Lemur*) has been more recent than the phyletic separation of the common ancestor of these two groups from the catarrhine type of monkey. Moreover these data demonstrate that, within the prosimian groups, there is no phylogenetic relationship between the Tupaioidea and the Lemuroidea.

A different immunochemical procedure has been used by Sarich and Wilson (1967). These investigators utilized a single protein fraction of the serum: the albumins. They claim that this molecule is relatively stable in any species and that the variation in the amino acid sequences which exists in different species is an index of the phylogenetic differences existing among them. They tested the homogeneity or evaluated the amount of heterogeneity of this serum fraction in several Primate species by means of the microcomplement fixation (MC′F) technique on antialbumin serum, obtained by immunizing rabbits with purified albumins of these Primates. In developing their research these scientists assumed, moreover, that the amino acid differentiation in the albumin molecules changes to similar degrees over a given period of time in different organisms. On the basis of this assumption they tried to quantify the phylogenetic relationships and particularly the times of divergency of various taxa as summarized below:

Divergence between	Time (millions of years)
Lemuriformes	*70–80* (assumed)
Lorisiformes/Simiae	
Tarsiformes	
Platyrrhines/Catarrhines	35–40
Cercopithecoidea/Hominoidea	20–24
Hylobatidae/Hominoidea	10–12
Pongo/other Hominoidea	9–11
Pan Gorilla /Homo	4–5

The most controversial aspect of these findings is the close relationship it suggests between man, chimpanzee and gorilla indicating that they all had a common ancestor very recently. Other scientists in fact, on the basis of other biological information suggest that albumin does not change at a constant rate, but its evolution may have been retarded in either apes or humans since their separation. Thus, although their albumins are undoubtedly very similar, this cannot be taken to mean that they have been evolving independently for only a short while. Sarich and Wilson, however, claim that this argument is totally invalid and that there is no evidence of a slowdown in the evolution rates in primates.

6. The Use of the Chromatographic Method in Taxonomic Research

As well as electrophoresis, a second method in common use in the study of serum proteins is chromatography. The chromatographic system takes its name from the fact that it was originally used to separate coloured compounds (chlorophyll *a* and *b*). It consists of a base functioning as a fixed phase on which migrate the different substances carried by various solvents that act as the mobile phase.

Essentially the migration occurs because of the affinity the mobile phase has for the fixed phase. The chromatographic mobility of a particular substance is defined with respect to a determinate solvent as the relationship between the distance travelled by the substance and the distance travelled by that solvent indicated by R_f.

The identification of the spots on the chromatogram is effected by using suitable substances to single out selectively various amino acids: ninhydrine (hydrate of triketohydrindene) for example, yields violet-coloured compounds specifically with amino acids (proline and oxyproline, yellow).

The first biological subjects studied were pathological individuals who excreted abnormal amino acids in their urine.

By comparing the patterns of amino acids in the urine of healthy and sick individuals it is also possible to determine the cause of the anomaly. For example, in the course of these studies it was discovered that the secretion of β–amino isobutyric acid, which appears in only some patients, is subject to genetic control and occurs only in cases of homozygosity of the gene (recessive).

These first results have led to a flowering of biochemical-genetic research at the level of the urinary amino acids and of the final stage of their metabolism (indoles and amidazoles). This kind of research has been undertaken first of all, to uncover possible variations of the genic frequencies within the human species (for example indoles decrease in frequency from east to west starting from China) and secondly, to point out resemblances between diverse species (Fig. 9.11).

Since the differences in concentration between the different amino acids are not at all or very slightly dependent on diet, it is possible on a preliminary basis to say that there are very noticeable differences between the

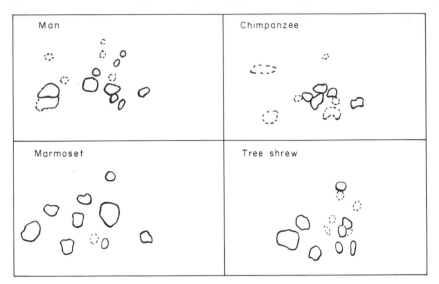

FIG. 9.11 Chromatograms of the urinary amino-acids of various primates (redrawn from Hawkes, 1968).

patterns of the anthropoid apes and those of man. In every case there are quantitative differences, and in some cases the qualitative ones, for all the differences in amino acids studied, are much more pronounced between the anthropoids and man than between the diverse species of anthropoids. Man excretes much more creatine in his urine than do the anthropoid apes, while the latter excrete a strong concentration of aspartic acid and glutamic acid. In this case their urine is more similar to that of human infants than to that of human adults. Smith has recently published some new research in this area (Smith, 1971).

GENERAL REFERENCES

Anfinsen, C. B. (1961). "The Molecular Basis of Evolution". John Wiley, New York.
Barnicot, N. A. (1969). Some biochemical and serological aspects of primate evolution. *Sci. Prog. Oxf.* **57**, 459–493.
Ceppellini, R. (1966). Genetica delle immunoglobuline. *Atti. Ass. Gen. Ital.* **12**, 1–129.
Conti Fuhrman, A. M. (1967). Elettroforesi bidimensionale su gel d'amido delle proteine sieriche delle Pongidae, delle Hylobatidae e di alcune specie di Cercopithecidae. *Arch. It. per l'Antr. e Etn.* **98** (3), 183–190.
Conti Fuhrman, A. M. and Chiarelli, B. (1968). Taxonomic and phylogenetic interest of the study of serum proteins of Old World Primates using bidimensional electrophoresis on starch-gel. *In* "Taxonomy and Phylogeny of Old World

Primates with References to the Origin of Man", Rosenberg and Sellier, Turin.

Crowle, A. J. (1961). "Immunodiffusion", Academic Press, New York and London.

Gartler, S. M., Firschein, I. L., Dobzhansky, I. (1956). A chromatografic investigation of urinary amino-acids in the great Apes. *Am. J. Phys. Anthrop.* **14**, 41–56.

Goodman, M. (1962). Evolution of the immunologic species specifity of human serum proteins. *Hum. Biol.* **34**, 104–120.

Goodman, M. (1963). Serological analysis of the systematics of recent Hominoids. *Hum. Biol.* **35**, 377–390.

Goodman, M. (1964). Man's Place in the Phylogeny of the Primates as Reflected in Serum Proteins. *In* "Classification and Human Evolution" (S. L. Washburn ed.), pp. 204–34, Methuen & Co. London.

Goodman, M. (1968). Phylogeny and Taxonomy of the Catarrhine Primates from Immunodiffusion Data. A review of the major findings. *In* "Taxonomy and Phylogeny of Old World Primates with References to the Origin of Man", Rosenberg and Sellier, Turin.

Hafleigh, A. S. and Williams, C. A. (1966). Antigenic correspondence of serum albumins among Primates. *Science N.Y.* **151**, 1530.

Jayle, M. F. *et al.* (1962). "Les Haptoglobines. Étude Biochimique, Génétique et Physiopathologique", Masson & Cie, Paris.

Kramp, P. (1956). Serologische Stamenbaumforschung. *In* "Primatologia", Vol. I (H. Hofer, A. H. Schultz, D. Starck, eds), S. Karger, Basel and N. York.

Sarich, V. M. and Wilson, A. C. (1967). Rates of albumin evolution in Primates. *Proc. natn. Acad. Sci. U.S.A.* **58**, 142–148.

Schultze, H. E. and Heremans, J. F. (1966). "Molecular Biology of Human Proteins", Vol. I, Elsevier Publishing Company, Amsterdam.

Smith, I. (1968). Indoles, Amino Acids and Imidazoles as an aid to Primate Taxonomy. *In* "Chemotaxonomy and Serotaxonomy", (J. G. Hawkes ed.), Academic Press, London and New York.

Smith, I. (1971). Comparative biochemistry of the Primates. *Folie Primatologia* **14**, 101–117.

IO

The Genetic Bases of Morphological Characters

1. A GENERAL REVIEW OF THE INHERITANCE OF MORPHOLOGICAL CHARACTERS

We have already seen that the shape of the nose itself can attest to relationships between us and our relatives. Many details of the human nose are determined by heredity, and thus appear to be more closely associated in related individuals than in individuals chosen at random. Because of the complex events that occur during the development of a gene to its eventual phenotypic expression, it is difficult to determine the number of the genes that form a particular nose or mouth. The ultimate manifestation of all morphological characters depends on a multitude of factors which influence the way in which they find expression.

The hereditary determinant of many quantitative morphological characters must, moreover, be sought not in simple Mendelian characters (major genes) like those which we have seen in the preceding chapters, but in complexes of multiple genes usually called "polygenes" whose action develops with an additive effect without complete dominance (pleiotropic effect).

Nevertheless, data of this type are not less important than those obtained for simple Mendelian characters, which we have seen in the earlier chapters. They make it possible to approach the study of the characteristics of the skeleton at the genetic level.

The skeleton, more than any other part of the body, can be used to demonstrate the evolution of the various species of primates because it is the single element to which reference can be made in comparisons with fossil remains. Indeed it must be remembered that the face, mandible, teeth and entire skeleton of man as he is today are the result of a long series of evolutionary stages which, through the line of antecedent reptiles and mammals, has led to the existing structure of the highly specialized primates. In examining the various features of the skeleton we must keep this clearly in mind, because some alterations may be repetitions of the characteristics of one of our ancestors who lived thousands or millions of years ago. Genes, as we have seen, do not disappear, but are transformed and eventually may reassume their ancestral form. Still, morphological characters are difficult to study because their genetic determinations interact among themselves in the expression of the phenotype, so producing an almost continuous variability. For this reason it is almost always necessary to consider "complexes of characters" as a unit, giving particular attention to their functional and adaptive significance, rather than to study single characters. This enormously complicates the type of analysis, and only the most modern biometric methods and the use of computers make progress on these problems.

In examining the skeletal characteristics of a primate species sexual differences must also be given special attention. The sex hormones influence the features of the skeletons of many species of primates in such a decisive manner that every bone may bear indications of the sex of the individual to whom it belonged. This, among other things, may be one of the reasons for the multiplication of specific and even generic names in palaeontology. An accurate revision of the fossil remains of the primates and hominids, keeping in mind this sexual difference, would certainly reduce by one third the number of species described.

Skeletal characteristics and dimensions are not the only features that are suitable to this type of analysis. Though they are difficult to study with the methods at present available, some characters determined by heredity, such as the size and weight of the body and its organs are receiving more attention. In all Mammals, the relationship between the weight of each single organ and the weight of the body follows the so-called allometric law. This law when applied to data of organ weight *vs* body weight plotted on log-log paper means that straight lines are yielded. In Primates, this law applies to all structures except the brain.

Other data which deserve further analysis are the different parameters of life: duration of embryonic life (gestation period), length of time required to reach sexual maturity, and the total duration of life.

The study of the morphology of male germ cells under the light and electron microscope reveals complexity that is still obscure in the primates. Their evolutionary history is almost completely independent from the events which caused the variation in other body structures, and, moreover, all of the genetic determinants of the male germ cell must be located in the Y chromosome, which gives them even more interest.

2. DERMATOGLYPHICS OF THE HAND

The papillary relief on the skin of the volar surfaces of the fingers, like the ridges and crease lines on the palm, are predetermined and fixed for every individual. They are already present in the skin during foetal life and if a piece of skin is removed from an adult it grows again in the same patterns. That they are determined by heredity in our species is proved by the very close correspondence of these characteristics in monozygotic twins and by genealogical studies. Another interesting proof of heredity is the concomitant finding of profound alterations in the palmar dermatoglyphs of individuals of our species in whom a supplementary twenty-first chromosome is present, Down's Syndrome.

The significance of these characters, the hereditary determinism of which is extremely complex, lies in the fact that they represent a general characteristic of all the various primates and present a wide range of variation among the different species while maintaining a degree of constancy at the species level.

Their functional reference is the relationship they bear to the prehensile function of the primate hand. The papillary ridges of the skin of the fingers and palm have a close relationship not only with the crease lines of the hand but also with the action of the limbs. They are fleshy cushions that protect the deep structures and serve to absorb the pressure when the skin is compressed against the bony parts of the phalanges or of the hand. They are completely free from hair and sebaceous glands but contain an enormous number of hypertrophied sweat glands and very many sensory nerve endings. In this way they form a complex mechanism not only for tactile sensation but also for increasing the adhesive effect of the touching surface, rather like an automobile tyre.

In contrast to the other species of mammals, all the living primates have hands and feet with particular dermatoglyphic characteristics which show significant differences among the various species, either as regards perfections or completeness. Most of the prosimians demonstrate these special dermal lines solely on their finger-tips while on their palms and soles they have simple fleshy ridges which are not always completely glabrous. The palms of the hands and the soles of the feet of almost all the monkeys, on the contrary, are covered by complicated dermatoglyphs. The patterns of the digital dermatoglyphics of the three groups of prosimians (Lemurs, Lorises and Tarsiers) differ markedly among themselves and in general from those of the monkeys. This lends support to the hypothesis that these three existing groups of prosimians represent three distinct lines of evolution, and that the monkeys are not particularly closely related to any of the three.

Within the group of platyrrhine monkeys, the Cebidae present dermatoglyphics that are remarkably variable in form and organization, while those of the Callithricidae are very stable and uniform for the entire family. Furthermore, *Callimico goeldii* is closely associated with the Callithricidae in this character. Among the catarrhine monkeys, the Cercopithecoidea are seen to be much more generalized and uniform than the Hominoidea. In the Cercopithecinae group, particularly, because of a wide interspecific variability, an exact delineation of the species on the basis of the characteristics of the dermatoglyphics is quite impossible. The dermatoglyphic characteristics in this case are specific for the group. In the Colobinae (*Nasalis* and *Colobus*) the specific differences are more marked. *Pygathrix* and *Presbytis* have dermatoglyphics that are differentiated only with difficulty from those of the Cercopithecinae. This relative uniformity of the characteristics of the skin of the extremities of the Cercopithecoidea is distinctive from the variability seen in the Ceboidea and Hominoidea (see Fig. 18.2, p. 287).

On the other hand the structural specializations of the dermatoglyphs of the Hominoidea are very marked. For this character too, the Hylobatidae are found to be completely distinct from the Pongidae and the Hominidae, with some characteristics that bring them closer to the Cercopithecidae. The Pongidae (*Pongo, Gorilla and Pan*) have many characters in common with our species, along with others that are unique to themselves.

3. SOME MORPHOLOGICAL CHARACTERISTICS OF THE SKELETON

The inheritance of morphological or rather morphometric characters in our species has been demonstrated by the comparison of monozygotic and dizygotic twins and by the study of some genealogies. The lengths of the cranial diameters and those of the face, for example, show a marked concordance in monozygotic twins, whereas dizygotic twins differ substantially from one another. This also applies, even if it cannot be demonstrated directly by statistical analysis (indices of correlation) to the shape of the cranium and of the head in general.

There is no reason to doubt that these and other morphological characters of the skeleton display a high degree of hereditability in the various species of primates as well. This is proved by the greater morphological similarity between the more closely related species. The difficulty lies, as we have said before, in discovering the correct character toward which to direct our attention, and in adopting an appropriate method of analysis.

A group of characters that lend themselves particularly well to this type of analysis are the so-called epigenetic characters of the skeleton. These are characters of a discontinous nature, the determination of which lies within the limits of presence or absence of relief features in the crania and other bones of the most diverse forms of vertebrates.

As for morphometric characters it is not clear whether the variations in the frequency of these characters and the characters themselves are controlled by one or more genes, even if for the most part it is thought that the genes which control them are of a reduced number. Nevertheless, the fact that they are found in different frequencies in diverse populations makes it possible to treat them as single hereditary characters and thus to evaluate the "distances" between one group and another.

There are, for example, the sulci on the external surface of the frontal bone that correspond to branches of the supraorbital nerves. The incidence of this trait varies from 0% among the Australian Aborigines to 50% among the Negroes. Other characters of this type in our species are the supra-orbital foramina, the infraorbital foramina, the direction and the type of the transverse palatine suture, the metopic suture, the orbital osteoporosis, the mental foramen, the presence of wormian bones, the mandibular torus, the palatine torus, the exostosis of the external canal of the ear etc. These are found in very diverse frequencies in human populations.

Some of these characters are also present in the various primate species with various frequencies; others are fixed characteristics found in all the individuals of a particular species. Still others, are completely absent from the skeletons of all the members of a species. An extensive study of these characters on the skeletons of the different primate species would certainly help us at least to distinguish the variable characters from the stable ones, thus contributing to a more adequate evaluation of the characters worth considering for phylogenetic studies of fossil remains of primates.

4. The Importance of the Teeth to Evolutionary Research

The comparison of the dentition of various living species with the corresponding fossil forms very often demonstrates better than the study of other parts of the skeleton, the transitionary forms and the relationships between species. The teeth of modern man, for instance differ from those of his fossil ancestors in their dimensions and morphology. Similar, but even greater differences are found between man and the anthropoid apes.

Another reason why the study of the teeth is important in clarifying the phylogenesis of the primates is the fact that the major part of the identifiable fossil remains of this group consists of teeth. In fact, they are better preserved than any other part of the skeleton during the process of decomposition, which for the primates is generally more or less complete because of their particular forest habitat.

The teeth of the primates have thus been studied in great detail and their morphological characteristics are decisive in the attribution of a fossil find to a determinate family or genus or even, sometimes, species. Obviously deductions based on the study of the teeth will tend to be the subject of any discussion on the phylogenetic relationships within this group.

Since they develop from the very first stage of life and during development are protected by the mandibular and maxillary bones the morphology of the teeth is not subject to any external environmental modification, and is influenced only by the genetic constitution of the individual. It is thus clear that, when properly studied they constitute a secure basis for the establishment of systematic and phylogenetic relationships. However, an adequate interpretation of the dental characteristics of the primates is by no means simple.

Firstly, the number of specimens of fossil primate teeth is rather small so that little can be said regarding individual variability. Very often the Palaeoprimatologists tend to undervalue the individual variability of characters. Knowledge of the variability of the dental characteristics of species actually living could be useful in the interpretation of the data on the fossils, but, save for a few exceptions, information on this point is also somewhat scarce.

Then, when comparisons are made on a wider level, as for example between different genera or different families, other problems arise. If two forms resemble each other are they really related, or are they the product of a parallel evolution? In this case, the probability cannot be calculated exactly but must be estimated; the more detailed the resemblance is, the less likely it is that the character in question would have evolved more than once. This type of estimation must not be overdone either; there are complicated morphological changes that may be due to simple and easily repeated genetic variations.

At one time, for example, it was believed that the "Apatamydae" of the lower Tertiary were related to the present aye-aye, which it resembles as regards dentition, particularly as they both have big incisors. However, the aye-aye is anatomically and physiologically very similar to the lemurs of Madagascar and it is reasonable to suppose that the resemblance in dentition to that of the Apatamydae is due to parallel evolution. In fact, large incisors are found in many other mammals belonging to different families.

On other occasions the similarities are indeed indications of a fairly close relationship. *Proconsul*, for example, exhibits dental characteristics very similar to those of *Pliopithecus*, characteristics for which both differ from the present anthropoids. *Pliopithecus* is generally considered to be one of the ancestors of the gibbons, while *Proconsul* is thought to be one of the ancestors of the present-day great apes. The degree of likeness at the level of dental characteristics is such as to make the division of the Hominoids of the Miocene into two families appear artificial. Probably the differentiation and separation must have come about at a later time.

Resemblance does not necessarily prove relationship, just as differences do not necessarily demonstrate the lack of it. It is largely a question of how the differences observed are interpreted in a hypothetical evolutionary process. The example of *Oreopithecus* is interesting in this regard. Initially Hürzelar, on the basis of various characters (especially dental ones like the small size of the canines, the absence of a diastema and the slight specialization of the premolars) considered *Oreopithecus* more closely related to the Hominids than to the Pongidae. From his conclusion the possibility was suggested that the Hominids had already become differentiated from the Pongidae before the beginning of the Pliocene.

Many differences both in the Pongidae and the Hominidae were found in the characteristics (including the dental ones) of *Oreopithecus*, and at the same time many characteristics in which the Pongidae and the Hominidae resembled each other but were differentiated from *Oreopithecus*, were discovered.

These data indicate that *Oreopithecus* evolved in a direction separate to that of both these families, and that it was separated from the Hominoids at a time preceding the differentiation of the Pongidae.

It is simpler, as we shall see, to derive the dental characteristics of man from those of *Proconsul* or even of *Pliopithecus* than from those of *Oreopithecus*.

Before going into details of the dentition of the various species of fossil and living primates we shall turn our attention to the principal dental characteristics of primates in general. Only the permanent teeth will be considered.

5. CHARACTERISTICS OF HUMAN DENTITION

The permanent dentition of modern man consists of 32 teeth, 16 in the maxilla and 16 in the mandible. Each side thus contains eight teeth of four distinct types, which differ among themselves in shape and in use: incisors, canines, premolars and molars. The formula of the permanent dentition of man is: $\dfrac{\text{2I, 1C, 2P, 3M}}{\text{2I, 1C, 2P, 3M}}$.

The incisors, because of their upper margin, are used chiefly to bite and cut. The upper central incisors are bigger than the upper lateral ones. The latter are sometimes very irregular in shape and size, and sometimes they are completely missing. The lower ones are small and very similar. Generally they close with the cutting edge of the lower ones slightly displaced towards the tongue. An occlusion (cutting edge of the lower ones against the cutting edge of the uppers) is frequent, particularly in populations which live on hard and fibrous foods.

The canines are placed at the sides of the mouth, and are usually pointed. In some mammals they constitute an excellent weapon of defence, in man their principal function is the cutting and the tearing of fibrous foods. Their length and position are such that they do not normally project beyond the level of the other teeth. This allows a lateral movement during mastication which is not possible in the species of primates equipped with long canines.

The teeth following the canines, in an anterior to posterior direction, are the premolars. The permanent premolars replace the deciduous molars, but assume a different function. Each of them has two cusps, although sometimes the second premolar has a third small cusp. The first lower premolar often is flattened, as if it had been compressed, and has a pointed prominence on the external margin which, occluding with the upper canine, gives cutting possibilities which must have been highly developed in our ancestors. The roots of the premolars of modern man are reduced in comparison with those of the fossil and living Pongidae. The first upper premolar generally has two roots but sometimes only one, the second usually has one root and very occasionally two; the lower premolars have one root each.

In man the molars are used almost exclusively for grinding. The upper molars have four (sometimes three) cusps, while the lower have five (sometimes four). Their number and position and the type of sulci between them differ from one individual to another but constitute, as we shall see, an important diagnostic character for each species. Generally the upper molars have three roots, the lower two. In European-type populations the first molar is usually large, the second smaller. In other populations the third is larger than the second, although smaller than the first.

6. The Evolution of Primate Dentition

There are several theories attempting to explain the evolution of the teeth from the simplest forms of the pointed teeth of the first reptiles to the most complicated forms of human molars.

The most accredited hypothesis is that the simple pointed tooth, was originally derived from a placoid scale, and later developed two small additional cone-shaped protuberances, one in front and the other behind the principal cone. This stage is known as the triconodont stage, and is represented by fossils of now-extinct mammals.

It is supposed that these two added structures (cones) moved externally on the upper teeth and internally on the lower ones to form a triangle. These trituberculate teeth must be considered as the basic ancestral form from which all the dentitions of the living mammals were originated by the addition, elimination or modification of cusps.

The different types of dentition of the various species of primates, like those of the other mammals, came into being under the pressure of selection and follow very definite biological laws. Some types of dentition are very limited and are the result of adaptation to a particular type of diet. The carnivores cannot chew the grass which constitutes the essential food of the horse, nor can sheep eat meat. Aside from these extreme specializations there are other groups of animals, including many primates and man, whose dentition is little specialized because for millenia they have been omnivorous. All the primates have a very similar dentition. The principal differences are found in the intensity of expression of certain characteristics, or in the relative sizes of the various parts, or in the number of teeth of each type (Table 10.1).

In general the teeth of the prosimians are coarser than those of the higher primates. They also exhibit a degree of variability, especially in the number of premolars, which may range from two to four for each quadrant. The aye-aye, because of its vegetarian specialization, has lost all four canines, the four lateral incisors and the lower premolars, and has retained only one of the upper premolars on each side.

The monkeys have a more constant formula for the number of teeth. As an exception to this, some of the New World monkeys (platyrrhines) have three premolars in each of the four quadrants, while the Old World monkeys (catarrhine) always have two premolars per quadrant. Consequently the standard dental formula of the catarrhine monkey is: $\dfrac{2I, 1C, 2P, 3M}{2I, 1C, 2P, 3M}$, which is fairly far removed from that of the permanent teeth of the primitive placental mammals (3 incisors, 1 canine, 4 premolars and 3 molars for each dental arch).

The molars of the catarrhine monkeys are almost uniform; they are quadritubercular and the peaks of the opposed cusps are furnished with prominent transverse crests. This type of molar, which is highly specialized, is called "bilophodont" (double-crested). Finally, in many species the last molar of the lower arch has one very large cusp.

In the different families of catarrhine monkeys (Cercopithecidae, Colobidae, and Hylobatidae) the various teeth assume special characters depending essentially on the diet to which they are adapted and on their phylogenetic distance.

Oreopithecus has a type of dentition which differs from both the Homi-

TABLE 10.1. Dental formulae of the living primates.

PRIMITIVE
PLACENTAL MAMMALS (total teeth 44) $I\frac{3}{3}, C\frac{1}{1}, P\frac{4}{4}, M\frac{3}{3}$

PROSIMII:

Tupaiidae
(total teeth 38) $I\frac{2}{3}, C\frac{1}{1}, P\frac{3}{3}, M\frac{3}{3}$

Lemuridae
(total teeth 36) $I\frac{2}{2}, C\frac{1}{1}, P\frac{3}{3}, M\frac{3}{3}$

except for   *Lepilemur*
(total teeth 32) $I\frac{0}{2}, C\frac{1}{1}, P\frac{3}{3}, M\frac{3}{3}$

Indriidae
(total teeth 30) $I\frac{2}{2}, C\frac{1}{0}, P\frac{2}{2}, M\frac{3}{3}$

Daubentoniidae
(total teeth 18) $I\frac{1}{1}, C\frac{0}{0}, P\frac{1}{0}, M\frac{3}{3}$

Lorisidae
(total teeth 36) $I\frac{2}{2}, C\frac{1}{1}, P\frac{3}{3}, M\frac{3}{3}$

Tarsiidae
(total teeth 34) $I\frac{2}{1}, C\frac{1}{1}, P\frac{3}{3}, M\frac{3}{3}$

PLATYRRHINES:

Callithricidae
(total teeth 32) $I\frac{2}{2}, C\frac{1}{1}, P\frac{3}{3}, M\frac{2}{2}$

Cebidae
(total teeth 36) $I\frac{2}{2}, C\frac{1}{1}, P\frac{3}{3}, M\frac{3}{3}$

CATARRHINES

Cercopithecidae
Hylobatidae
(total teeth 32) $I\frac{2}{2}, C\frac{1}{1}, P\frac{2}{2}, M\frac{3}{3}$

Pongidae
(total teeth 32) $I\frac{2}{2}, C\frac{1}{1}, P\frac{2}{2}, M\frac{3}{3}$

Hominidae
(total teeth 32) $I\frac{2}{2}, C\frac{1}{1}, P\frac{2}{2}, M\frac{3}{3}$

noids and the living catarrhine monkeys. Among other things the pos-
terior-anterior cusp of their upper molar is joined by means of a crest to
the oblique crest. The first and second lower molars have four main cusps
and a rudimentary hypoconid.

7. The Teeth of the Pongidae and of the Hominidae

A major feature which distinguishes the upper arch of the Pongidae from that of the Hominidae is the presence in the former of a space between the canine and the premolar, which is missing in the Hominidae. This space is called a diastema (Fig. 10.1).

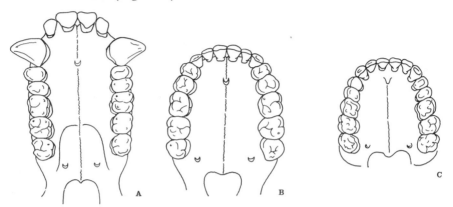

Fig. 10.1 Palate and upper teeth of A, *Gorilla* B, *Australopithecus* and C, Australian aborigine. The dentition of the Australopithecines is a reconstruction from remains of several individuals. Noteworthy are the relatively small dimensions of the canines and incisors, and as well the curvature of the dental arch.

In the Pongidae the alveoli of the upper central incisor are very large and are placed far forward in relation to the canines. Man's incisors, on the other hand are small and the posterior margins of their alveoli are at the same level, or behind the line joining the anterior margin of the canine alveoli.

The canines of the Pongidae are relatively large and rather pointed and when they close they are superimposed by a wide margin, and extend far beyond the plane of occlusion. This does not happen in modern man. In the Pongidae the canines erupt after the second molar and sometimes even after the third, while in modern man they normally erupt before the second molar. Moreover in the Pongidae these teeth show a pronounced sexual dimorphism, while in man this is not so.

Every discussion of the origin of man invariably introduces the problem of the canines. It appears that the most ancient form of the Pongidae, *Proconsul* of the African Oligocene, also had very small canines. This fact has led some scholars to conclude that the reduced canines of man are an indication of primitiveness. As a result of the specialization of the canines and their enlargement, the first premolar of the Pongidae has assumed a monocuspid form; in man this tooth is bicuspid.

The premolars have also remarkable diagnostic value for distinguishing a Pongid from a Hominid. In fact, in the Pongidae the first upper premolar normally has three roots, while in present-day man it has only one or two roots; only occasionally in some human populations are three roots present.

The most interesting teeth for comparative study are the molars. Normally, the molars are numbered from front to back, so that M_3 is the most posterior. A fourth, supernumerary, molar is found frequently in the orangutan, and only rarely in man. It is produced by a precocious division of the dental bud of the last molar (M_3).

As far as size is concerned, in the Pongidae the largest molars are usually the second in the upper and in the lower arch, while in modern man the first in both arches are the largest.

The upper molars have four main cusps, of which the two external ones and the anterior of the internal ones correspond to the three cusps of the basic type, the so-called *"trigone"*. The fourth cusp often remains joined to the transverse crest which in the Hominoidea (Pongidae and Hominidae) invariabily ends on the top of the fourth cusp. The posterior-interior cusp (*hypocone*) rises from the basal *cingulum* which encircles the *trigone*. It is usually the least prominent of the four.

The lower premolars of the Hominoidea have five undulating cusps: two internal and three external, which are distinguished from their homologues of the upper molars by the suffix "ide". Only the two anterior cusps project farther than the original trigonid and are often united by a bridge.

Both the Pongidae and the Hominidae exhibit a remarkable variability in the disposition of their cusps and in their relative and absolute dimensions. Comparing the general morphology of the cusps and crests of the molars of a Pongidae with those of a Hominidae, it is seen that on the whole they resemble each other in appearance, being slightly differentiated by the formation of the cusps and by the crests present on the occlusal surface. These crests are more prominent in the gorilla; the orangutan has a typically low and flat crown, covered with a great many small wrinkles in the enamel. In man the cusps, especially in the second and third upper molars, are often reduced to three, and the five cusps of the interior molars to four. The Australian aborigines nevertheless have five cusps on all three molars of the upper arch. Between the two posterior cusps there is often a minute sixth cusp (tuberculum-sextum). In the original lower molar the anterior-interior cusp (the metaconid) is fused with the central external cusp (hypoconid).

This outline is specific for all the Hominoidea. The most ancient example of such a fusion was found in the Miocene remains of Dryopithecine. It is expressed by the formula "Y5" which indicates the presence of five cusps between which a system of oblique and transverse sulci gives origin to a central Y. The Y5 arrangement is formed by the contact between the mesio-lingual and disto-buccal cusps, or sometimes by the mesio-buccal cusps, and disto-lingual cusps. It develops little by little into an arrangement with four cusps, indicated by "+4". The first inferior molar of modern man still shows traces of the "Y5" arrangement; the second on the contrary is quadritubercular.

GENERAL REFERENCES

Alciati, G. (1960). Sulla metodologia per lo studio dei rilievi cutanei palmari. *Riv. Antrop.* **47**, 89–127.

Biegert, J. (1964). The evaluation of characteristics of the skull, hands, and feet for primate taxonomy. *In* "Classification and Human Evolution", S. L. Washburn (ed.), pp. 116–145, Methuen & Co., London.

Biegert, J. (1971). "Dermatoglyphics in the Chimpanzee. The Chimpanzee" Vol. 4, pp. 273–324 (J. Borne, ed.), Karger, Basel.

Billsborough, A. (1969). Rates of evolutionary change in the hominid dentition. *Nature Lond.* **223**, 146–149.

Chiarelli, B. (1967a). Lamorphologia del chromosoma Y delle differenti specie di Primati. *Riv Antrop* **54**, 137–140.

Chiarelli, B. (1967b). Longerita e periodo di gestazione nei Primati non-umani. *Riv. Antrop.* **54**, 159–160

James, W. W. (1960). "The Jaws and Teeth of Primates", Pitman Medical Publishing Company, London.

Mavalwala, J. (1971). The heredity of dermatoglyphic traits in Non-Human Primates and Man *In* "Comparative Genetics in Monkeys, Apes and Man (B. Chiarelli ed.), Academic Press, London and New York.

Penrose, L. S. (1970). I dermatoglifi. *Le Scienze*, 56–64.

Schuman, E. L. and Brace, C. L. (1955). Metric and morphologic variations in the dentition of the liberian chimpanzee; comparisons with anthropoid and human dentitions. *In* "The Non-Human Primates and Human Evolution", J. A. Gavan (ed.), pp. 61–90, Wayne University Press, Detroit, Mich.

Schulz, A. H. (1970). The comparative uniformity of the Cercopithecoidae, *In* "Old World Monkeys: Evolution, Systematics and Behavior" (J. R. Napier and P. H. Napier, eds), p. 39, Academic Press, New York and London.

II

The Karyotype of the Primates and of Man

1. GENERAL

We have seen in the second chapter that chromosomal variations, concomitant with special ecological isolation and with the succession of generations, have certainly played an important role in the diaspora of the primate species and in human evolution. Until a few years ago a detailed study of mammalian chromosomes was an extremely arduous undertaking and there was marked lack of precision, even for the human chromosomes. At present the methods of culture *in vitro* and the techniques of hypotonic treatment make it possible to study the somatic chromosomes at a satisfactorily accurate level from the point of view of both measurement and morphology. This has contributed to an objective comparison of the chromosomes of the various species and made possible deductions regarding phylogenesis.

In fact, although it is not possible to study the karyotype of the ancestors of living species, the study of the karyotype of the latter furnished many examples of the stages through which karyotypes have passed during their evolution. The similarities between the chromosomal complements of different but related species can be taken as an indication of common origin, while the differences make practicable a rough reconstruction of the stages by which they evolved.

Nevertheless, it is essential to point out that a likeness between the chromosomes of related species does not necessarily imply that they are homologous. Only the study of the meiotic chromosomes of interspecific hybrids will permit the establishment of possible homologies between apparently similar chromosomal complements. For the primates this type of analysis has not yet been done at all, and consequently the phylogenetic relationship which we deduce on the basis of the similarity of their karyotypes are anything but proved. Even with these limitations, the data collected on them can well be used to organize a phylogenetic system using the chromosomes as a foundation. Such a system, although still incomplete, provides many stimulating ideas for further research.

2. THE CHROMOSOMES OF THE PROSIMIANS

A synthesis of the available information on the chromosomes of the different species of Prosimiae is presented in Table 11.1. Data on the DNA content are reported in Table 11.3.

Within the Tupaioidea the chromosomes of *Tupaia glis* ($2n = 60$–62), *T. montana* ($2n = 52$–68), *T. minor* ($2n = 66$) and *Urogale everetti* ($2n = 44$) have been studied.

Perodicticus potto has 62 chromosomes of which 24 are submetacentric and 36 acrocentric, apart from the sex chromosomes.

Nycticebus coucang and *N. pygmaeus* both have 50 chromosomes, all submetacentric. *Arctocebus calabrensis*, on the contrary, has 52 chromosomes although its karyotype is very similar to that of *Nycticebus*. *Loris tardigradus* has 62 chromosomes some of which are submetacentric and some acrocentric.

Among the Galagidae, *Galago senegalensis* and *Galago crassicaudatus* have been studied. The first has been found to have a chromosome number of $2n=38$, and the second one of $2n=62$. The chromosomal complement of both these species have been studied in detail by various authors who have found different proportions of submetacentric and acrocentric chromosomes. It is therefore probably correct to think that it is a question of species which represent a chromosomal polymorphism, but it is difficult to explain the great difference between the chromosome numbers of the two species. The diploid number of chromosomes of the Lemuroidea ranges between $2n = 44$ and $2n = 66$. Within the limits of the subfamily Cheirogaleinae only two species have been studied thus for, one for each genus.

Among the Indriidae only the number of chromosomes of *Propithecus verreauxi* ($2n=48$) is known. No morphological details have been published.

Among the Tarsioidea the chromosomes of *Tarsius syrichta* and *Tarsius bancanus*, both with $2n=80$ chromosomes, have been studied. The chromosome complement of *T.bancanus* is made up of 14 submetacentric and 66 acrocentric chromosomes.

TABLE 11.1. Numerical data on the chromosomes of the Tupaiidae, Lorisidae, Galagidae, Lemuridae, Indridae Daubentoniidae, and Tarsiidae.

Taxa	2n	S–M	A	X	Y
Tupaiidae					
Tupaia glis	60–62	14–12	44–48	M–S	A
Tupaia montana	52–68			M	A
Tupaia minor	66				
Urogale everetti	44				
Lorisidae					
Loris tardigradus	62	34–38	26–22	S	S–A
Nycticebus coucang	50–(52)	48		S	S
Nycticebus pygmaeus	50				
Arctocebus calabrensis	52	50		S	
Perodicticus potto	62	24	36	S	A
Galagidae					
Galago senegalensis	38	22–24–30	14–12–6	S	S–A
Galago crassicaudatus	62	6–30	54–30	S	A–S
Galago alleni	40				
Galago demidovii	58				
Lemuridae					
Microcebus murinus	66		64	A	A
Cheirogaleus major	66		64	A	A
Phaner sp.	48				
Hapalemur griseus	54–58	10–6	42–50	A	A
Lemur catta	56	10–14	44–50	S–A	A
Lemur variegatus	46	18	26	S	A
Lemur macaco	44–48–52–58–60	20–16–4	22–30–52–54	A	A
Lemur mongoz	58–60	4	52–54	A	A
Lepilemur sp.	22–38				
Indridae					
Propithecus verreauxi	48				
Daubentoniidae					
Daubentonia madagascariensis	30				
Tarsiidae					
Tarsius syrichta	80				
Tarsius bancanus	80	14	66		

Although intergeneric chromosomal variation is of great interest, a more general question arises as to whether phylogenetic relationships between the families of the Primates can be established by looking at the variations in chromosome numbers.

The data on diploid numbers of somatic chromosomes do not give any information about frequent chromosome mutations such as centric fusion. This kind of information can be deduced by applying Matthey's fundamental number (F.N.) which takes into consideration the number of arms only, and is based on the assumption that centric fusion (or misdivision of the centromere) is one of the more successful mutations to pass through the sieve of meiosis. This kind of mutation, moreover, is one which would not interfere with the organization of the genetic information on the chromosome, but would only change, by reduction or increase, the random distribution of genetic information in the offspring.

A reduction or increase in chromosome units would, however, present an adaptive advantage to an organism because it would reduce or increase the potential variability within a population. A relationship between chromosome morphology and chiasma frequencies must surely exist, although at present there is not enough information to define exactly what form it may take.

The fundamental number (F.N.) may be calculated by counting metacentric and submetacentric chromosomes as 2, and acrocentric and subacrocentric chromosomes as 1 in a general karyotype. The karyological information elaborated in this way yields more reliable taxonomic results.

The F.N. in Tupaioidea ranges between 70 and 84; in Lorisidae, between 87 and 102; in Galagidae, between 61 and 94; in Lemuridae, between 62 and 70. The F.N. in Tarsiidae is 94, and it is 54 in Daubentoniidae. These two latter families appear to be quite distinct both in comparison to each other and to other prosimians.

The taxonomic, and hence phylogenetic separation between Tupaioidea with respect to Lorisoidea and Lemuroidea does not present any difficulties. Between the Lorisoidea and the Lemuroidea, however, difficulties do arise, and a conflict of opinion exists about their separation. The diploid chromosome numbers are completely overlapping, so they cannot be used to distinguish the two superfamilies, but data on the F.N. show that they are distinctly separate and not superimposable because Lorisidae have F.N. 87 to 102, and Lemuridae have F.N. 64 to 70.

The taxonomic position of the Galagidae whose F.N. represents in some way a bridge between the Lemuridae and Lorisidae (F.N. varying between 61 and 94) is of particular interest.

The strikingly large variation in the diploid chromosome numbers in the two *Galago* species (*Galago senegalensis*, $2n=38$, and *G. crassicaudatus*

$2n=62$) suggested the possibility of a polyploid mechanism in the origin of this variation.

However, the established chromosomal polymorphism in the chromosome number of *Galago senegalensis* is due to the mechanism of centric fusion, and the diploid chromosomal number is 58 in *Galago demidovii* and 40 in *G. alleni*. These data and the establishment of identical DNA content in the two *Galago* species ($7 \cdot 54 \pm 0 \cdot 17$ in *G. senegalensis* and $7 \cdot 26 \pm 0 \cdot 09$ in *G. crassicaudatus* in a.u.) have eliminated the hypothesis of polyploidy. Moreover, they give specific support to a stricter taxonomic relationship between the Galagidae and the Lorisidae and open the field to extensive speculation about the adaptive advantages of chromosomal polymorphism in *Galago*.

The fact that an extensive polymorphism existing in the genus *Lemur* is due to a centric fusion mechanism and even produces karyological subspecies, enormously increases the interest in the adaptive advantage of such variations and in such a peculiar mechanism of speciation.

Another aspect of interest concerns the relation of the living Prosimians with other groups of monkeys, especially the South American ones. According to a classic theory, the South American Monkeys originated from a group of extinct North American Prosimians (Omomydae) whose representatives migrated from North to South America. However, the recent revision of the theory of continental drift makes it possible to speculate that a direct migration of some early Primates came directly from Africa to South America up to the end of the Eocene. This stimulating hypothesis has to be tested utilizing different biological information of which chromosomes could be one.

3. THE CHROMOSOMES OF THE PLATYRRHINE MONKEYS

The diploid number of chromosomes of the platyrrhine monkeys studied to date varies from $2n=20$ to $2n=62$ (see Table 11.2).

The various species studied within the family Callithricidae present a natural uniformity in the number (from 44 to 46) and in the morphology of the chromosomes. Therefore they represent a particularly closeknit group of species. The variations which have led to the diversification of the number of chromosomes can all be brought back to mechanisms of centric fusion.

The family Cebidae includes ten genera, six of which have been studied in detail. *Aotus trivirgatus* has 50 to 54 chromosomes. Regarding morphology the karyotype is very similar to that of *Cebus*, *Alouatta caraya*, *C. acajao* and *Callicebus moloch*.

Two species of the genus *Callicebus* have been studied: *C. moloch* and

TABLE 11.2. Numerical data on the chromosomes of the Platyrrhine primates.

Taxa	2n	M	S	A	X	Y
Callithricidae						
Callithrix chrysoleuca	45	4	26	14	S	S
C. jacchus	46	4	28	12	S	A
C. argentata	44	4	28	10	S	M
C. humeralifer	44	4	28	10	S	A
Cebuella pygmaea	44	4	28	10	S	A
Saguinus oedipus	46	4	26	14	S	M
S. fuscicollis	46	4	26	14	S	M
S. nigricollis	46	4	26	14	S	M
S. mystax	46	4	26	14	S	M
S. leucopus	46	–	–	–	–	–
S. tamarin	46	–	–	–	–	–
Leontideus rosalia	46	4	28	12	S	M
Callimiconidae						
Callimico goeldii	48	4	24	18	S	A
Cebidae						
Alouatta seniculus	44	4	12	26	A	S
A. villosa	53	–	–	–	–	–
A. caraya	52	4	16	30	S	A
Aotus trivirgatus	54 (52)	4	16	32	S	S
Cebus albifrons	54	4	16	32	S	A
C. capucinus	54	4	14	34	S	A
C. apella	54	4	20	28	A	A
Callicebus moloch	46	4	16	24	S	A,M
Callicebus torquatus	20	–	10	10	–	–
Cacajao rubicundus	46	4	16	24	S	–
Pithecia pithecia	46	–	–	–	–	–
Saimiri sciureus	44	4	26	12	S	A
Saimiri madeirae	44	4	28	10	S	A
Saimiri boliviensis	44	4	28	10	S	A
Lagothrix ubericola	62	–	–	–	–	–
Ateles arachnoides	34	–	–	–	–	–
A. paniscus	34	–	30	2	S	A
A. belzebuth	34	–	30	2	S	A
A. geoffroyi	34	–	30	2	S	A,S

C. torquatus. The first (*C. moloch*) has 46 chromosomes of which 4 are metacentric, 16 submetacentric and 24 acrocentric, not including the sex chromosomes. On the other hand, a chromosome number of $2n=20$, which seems surprisingly low, has been found for the second (*C. torquatus*). It is possible that mechanisms of isolation and selection have, in a relatively brief time, produced this reduction in the number of chromosomes, which is the lowest thus far found in the primates and one of the lowest in the entire group of mammals.

The number of chromosomes of *Pithecia pithecia* is $2n=46$. There is no information regarding its karyotype. No data are available for the genus *Chiropotes*.

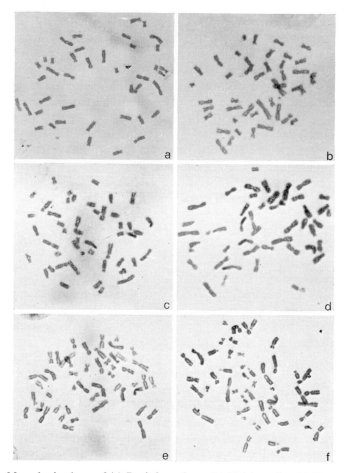

FIG. 11.1 Metaphasic plates of (a) *Papio hamadryas*, (b) *Hylobates lar*, (c) *Pongo pygmaeus* (d) *Gorilla gorilla* (e) *Pan troglodytes*, (f) *Homo sapiens*.

Cacajao rubicundus also has 2*n*=46 chromosomes, with a karyotype very similar to that of *Callicebus moloch*.

We have data for three species of *Alouatta*: *A. seniculus*, *A. caraya* and *A. villosa*. While *A. seniculus* has 2*n*=44 chromosomes, for *A. caraya* and *A. villosa* the chromosome numbers are 2*n*=52 and 2*n*=54.

All the individuals of the various species of *Cebus* studied, with the exception of one individual in four of *Cebus capucinus*, present 54 chromosomes with very minor morphological variations in their karyotypes.

The three species of *Saimiri* studied thus far (*S. sciureus S. madeirae* and *S. boliviensis*) all have a complement of 44 chromosomes, with4 metacentric, 26 submetacentric and 12 acrocentric, excluding the sex chromosomes. Some variation in different species seems, however, to exist.

Lagothrix has 62 chromosomes.

All the species of *Ateles* studied have 34 chromosomes, with no metacentric ones.

In general, the relationship between the total number of chromosomes and the number of acrocentric chromosomes is valid for the group of the Cebidae as well. This fact demonstrates that the Robertsonian mechanism for the reduction of the number of chromosomes has played an important role in the evolution and differentiation of the Cebidae.

Furthermore it seems that there exists in this group a direct relationship between the number of chromosomes and the degree of specialization for arboreal life. In fact the less specialized species, like those belonging to the genus *Cebus*, have in general a greater number of chromosomes and especially of acrocentric chromosomes, while those with a smaller number of chromosomes and few acrocentric ones, as for example *Ateles*, are better adapted to life in the forest.

Callimico goeldii presents 48 chromosomes of which 4 are metacentric, 24 submetacentric and 18 acrocentric, apart from the sex chromosomes.

In general the chromosomes of Cebidae differ greatly in number and morphology from the chromosomes of the Callithricidae. Nevertheless, *Callimico goeldii* provides an obvious example of a karyotype having characteristics intermediate between the two groups.

The taxonomic position of *Callimico* is still much discussed and there are doubts as to whether it should be classified among the Callithricidae or Cebidae. Its limbs and overall appearance would tend to classify it as a *Saguinus*, but its dental formula would place it among the Cebidae. From a karyological point of view *Callimico*, in general, appears to be more similar to a *Callithrix* although still presenting many resemblances to the karyotypes of *Cebus*, *Callicebus*, *Aotus* and *Cacajao*.

Very probably *Callimico* is a form which separated from the phylum of platyrrhine monkeys before they were differentiated and specialized, and has retained the karyological characteristics of the ancestral type.

The data relating to the DNA content of the nuclei bring to light significant differences between *Cebus* and *Ateles,* between *Cebus* and *Cebuella,* and between *Ateles* and *Alouatta.* These data, although demonstrating convincing taxonomic differences, require a more extensive analysis if they are to be interpreted from a phylogenetic point of view (Table 11.3).

TABLE 11.3. Nuclear DNA content and area of Primate lymphocytes (from M. G. Manfredi-Romanini, 1972)

Species	DNA ± s.d. (arbitrary units)	DNA (pg)
Lemur catta	12·18 ± 0·11	5·91
Lemur fulvus	12·80 ± 0·16	6·21
Lemur macaco	12·68 ± 0·16	6·15
Lemur mongoz	12·07 ± 0·10	5·85
Galago senegalensis	13·97 ± 0·09	6·78
Tarsius syrichta	18·95 ± 0·77	9·19
Cebuella pygmaea	12·31 ± 0·12	5·97
Cebus albifrons	14·07 ± 0·13	6·83
Cebus apella	13·01 ± 4·19	6·34
Cebus capucinus	13·45 ± 4·27	6·52
Cebus nigrovittatus	12·02 ± 0·19	5·83
Saimiri sciureus	11·66 ± 0·13	5·65
Alouatta caraya	13·59 ± 3·05	6·59
Alouatta palliata	12·82 ± 0·70	6·22
Ateles belzebuth	12·82 ± 0·15	6·22
Ateles geoffroy	11·48 ± 0·12	5·57
Ateles paniscus	12·26 ± 0·20	5·95
Cercopithecus aethiops	10·42 ± 0·35	5·05
Cercopithecus cephus	12·49 ± 0·57	6·06
Cercopithecus patas	12·43 ± 0·29	6·03
Cercocebus galeritus	16·72 ± 0·29	8·11
Cercocebus torquatus	17·31 ± 0·29	8·40
Macaca mulatta	11·11 ± 0·26	5·39
Macaca silenus	11·24 ± 0·18	5·45
Papio hamadryas	12·48 ± 0·48	6·05
Colobus polykomos	12·75 ± 0·13	6·18
Nasalis larvatus	15·27 ± 0·19	7·41
Hylobates agilis	9·80 ± 0·12	4·75
Hylobates lar	10·35 ± 0·16	5·02
Symphalangus syndactylus	10·54 ± 0·22	5·11
Pongo pygmaeus	14·50 ± 8·40	7·03
Pan troglodytes	13·60 ± 4·60	6·60
Gorilla gorilla	12·62 ± 7·30	6·12
Homo sapiens	12·36 ± 0·11	6·00

s.d. = standard deviation.

4. THE CHROMOSOMES OF THE CATARRHINE MONKEYS

The species of greatest interest in the level of evolution of karyotypes, because they are most directly related to man, are the catarrhine monkeys. Consequently we shall give particular attention to the chromosomes of these species.

In the description of the karyotype of a species the morphology and sometimes also the measurement of the total length of the chromosomes contained in the metaphasic plates of the different species have particular importance as well as the number of chromosomes. In fact these data can give us an idea of the quantity of genetic information contained in the different species.

The data available for each genus of catarrhine monkey are synthesized in Table 11.4. The number of chromosomes in this group varies from $2n=42$ to $2n=72$, with the highest frequency for the number 42.

All the species of the genus *Macaca* present a diploid number of chromosomes equal to 42. The differences between the karyotypes of the different species are of little importance and can be reconciled with structural rearrangements of the inversion or translocation types. The same chromosome number, $2n=42$, is characteristic of the genera *Papio*, *Theropithecus* and *Cercocebus*. In the genus *Papio* there are small differences between the chromosomes of the diverse species which, as in the case of *Macaca*, can be traced back to structural variations such as translocations and inversions. The same is true for the different species of the genus *Cercocebus*.

The genus *Theropithecus*, represented by the single species *T. gelada*, has a karyotype which morphologically is very similar to those of the various species of the genus *Papio* (Fig. 11.2).

In all the species of the genera, *Macaca*, *Papio* *Cercocebus* and *Theropithecus* the X chromosome appears to be of average size with the centromere more or less in the middle. The Y chromosome is very small, and when it is possible to distinguish the centromere, it seems to be metacentric.

A characteristic common to all these species, moreover, is that of having one chromosome of medium size with an achromatic region on one arm surmounted by a linear satellite. This shared endowment of chromosomes lends support to the theory that these four genera of the family Cercopithecidae had a common origin. Later geographical and ecological isolation would have led to the differentiation of the karyotypes and to the formation of groups of individuals, reproductively isolated from the others, who were to become the originating ancestors of so many species. In this regard, if a comparative analysis is carried out on the morphology of each single chromosome, it is found that a closer similarity exists between the different species of these genera. The resemblance is close between *Macaca* and *Papio* and *Papio* and *Theropithecus* than between *Macaca* and *Cercocebus*.

TABLE 11.4. Numerical data on the chromosomes of the Catarrhine primates.

Taxa	2n	S	A	X	Y
Cercopithecidae					
Papinae					
Macaca mulatta	42	40	–	S	S
M. sylvana	42	40	–	S	S
M. speciosa	42	40	–	S	S
M. fuscata	42	40	–	S	S
M. assamensis	42	40	–	S	S
M. silenus	42	40	–	S	S
M. nemestrina	42	40	–	S	S
M. radiata	42	40	–	S	S
M. sinica	42	40	–	S	S
M. irus	42	40	–	S	S
M. cyclopis	42	40	–	S	S
M. maura	42	40	–	S	S
M. (= Cynopithecus) niger	42	40	–	S	S
Papio sphinx	42	40	–	S	S
P. hamadryas	42	40	–	S	S
P. cynocephalus	42	40	–	S	S
Theropithecus gelada	42	40	–	S	S
Cercocebus albigena	42	40	–	S	S
C. aterrimus	42	40	–	S	S
C. galeritus	42	40	–	S	S
C. torquatus	42	40	–	S	S
Cercopithecinae					
Cercopithecus patas	54	38	14	S	A,S
C. talapoin	54	38	14	S	A
C. diana	58–60	–	–	–	–
C. d. roloway	60	–	–	–	–
C. nigroviridis	60	36	22	S	A
C. neglectus	(60)	–	–	–	–
C. cephus	66	48	16	S	–
C. mona	66	48	16	S	S,A
C. ascanius	66	–	–	–	–
C. petaurista	66	–	–	–	–
C. nictitans	66 (70)	48	16	S	–
C. l'hoesti	72	48	22	S	–
C. mitis	72	48	22	S	S,A
Colobidae					
Presbytis entellus	44	40	2	S	A
P. obscurus	44	40	2	S	A
P. senex	44	40	2	S	A
Colobus polykomos	44	42	–	S	–
C. kirkii	44	42	–	S	A
Nasalis larvatus	48	46	–	S	–

Table 11.4 (*continued*)

Taxa	2n	S	A	X	Y
Hylobatidae					
Hylobates lar	44	42	–	S	S
H. agilis	44	42	–	S	S
H. moloch	44	42	–	S	S
H. noolock	44	42	–	S	S
H. (*Nomascus*) *concolor*	52	44	6	S	S
Symphalangus syndactylus	50	46	2	S	S
Pongidae					
Pongo pygmaeus	48	26	20	S	S
Gorilla gorilla gorilla	48	30	16	S	S
Pan paniscus	48	34	12	S	A
P. troglodytes	48	34	12	S	A
Hominidae					
Homo sapiens	46	34	10	S	A

A classification based on these likenesses accords perfectly with the system based on morphology.

These karyological similarities are thus to some extent validated by the idioplasmatic continuity still evident among the various species of this genus. It is noteworthy that hybridological data brings to light a remarkable frequency of chance hybrids between individuals of different species of this group, whereas this does not happen among the others (see Fig. 11.3).

The species of the genus *Cercopithecus*, among which *Erythrocebus* must also be included, present a variable number of chromosomes ranging between 54 and 72 and more than one chromosome number may sometimes be found in the same species (*C. neglectus, C. nictitans*).

It might be thought that this numerical variability of the chromosomes could be due to mechanisms of centric fusion, starting with a species having a higher chromosome number and happening repeatedly, as has already been demonstrated with regard to some prosimians. Still, these hypotheses, to be justified, require the following conditions: (a) that the number of acrocentric chromosomes be in proportion to the total number of chromosomes; (b) that the total length of the genome be kept constant. In fact neither of these conditions appear to be satisfied. The total number of chromosomes is not absolutely in relation to the acrocentric chromosomes, and the total length of the genome increases in direct proportion to the increase in the number of chromosomes.

Furthermore, the presence of only one pair of chromosomes with a large achromatic region in all the species of the genus *Cercopithecus*, as in all the

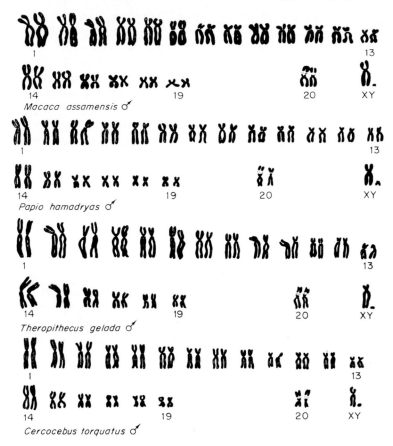

FIG. 11.2 Karyotypes of (a) *Macaca assamensis* (♂); (b) *Papio hamadryas* (♂); (c) *Theropithecus gelada* (♂); (d) *Cercocebus torquatus* (♂).

other catarrhines (except the anthropoid apes and man) contradicts the postulate that the phenomenon of polyploidy arose in an ancestor common to all these species. At present the hypothesis that phenomena of reduplication of some chromosomes occurred from time to time but never involved the chromosome having the achromatic region seems more plausible. The detailed morphological analysis of the somatic chromosomes and to an even greater degree the study of the meiotic chromosomes of these species should clarify the manner of origin of these variations in this group of primates, and should throw light on the still controversial chronology of their evolution.

They are undoubtedly of great interest as they represent an example of wide chromosomal variation in a relatively homogeneous taxonomic group.

The recent description by De Boer (1970) of a *Cerocopithecus nigroviridis* with $2n=48$ chromosomes gives this variation wider application
suggesting that the mechanisms of ecological and ethological isolation in a
forest habitat are very active in producing biological differentiation.

The data on *Cercopithecus* may also help in the interpretation of the
mechanisms which resulted in the differentiation and stabilization of the
number and morphology of the chromosomes in a species. Research in this
direction must, however, be developed before many of these species are
exterminated.

No data are available for the genera *Pygathrix*, *Rhinopithecus* and *Simias*;
species which are now extremely rare, but which could furnish important
karyological information.

The karyotype of *Colobus polykomos*, $2n=44$, appears to have many

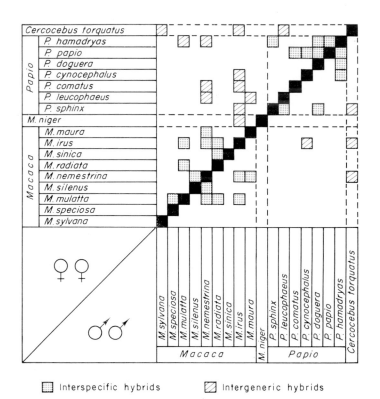

FIG.11.3 Graphical representation of the data on hybridization in the Cercopithecidae
found in the literature.

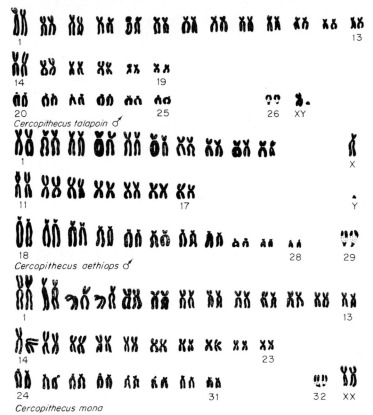

FIG. 11.4 Karyotypes of (a) *Cercopithecus talapoin* (♂); (b) *C. aethiops* (♂); (c) *C. mona* (♀) ssp. *campbelli*.

characteristics in common with that of *Presbytis* (*P. obscurus*), presenting equal numbers of diverse and morphologically similar chromosomes (Fig. 11.5).

The genus *Nasalis* has a diploid number of $2n=48$ chromosomes, the morphology of these being very similar to that of the chromosomes of the genera *Colobus* and *Presbytis*.

All the species of the genus *Hylobates* studied to date present a diploid number of $2n=44$ chromosomes, and among them there are absolutely similar chromosomes. The genus *Symphalangus*, on the other hand, displays a diploid number of $2n=50$ chromosomes, in general morphology fairly different from those of *Hylobates*. The chromosomes of the genera *Hylobates* and *Symphalangus* appear to have many morphological characteristics in common with those of the species *Colobus polykomos*, *Presbytis obscurus* and *Nasalis larvatus*.

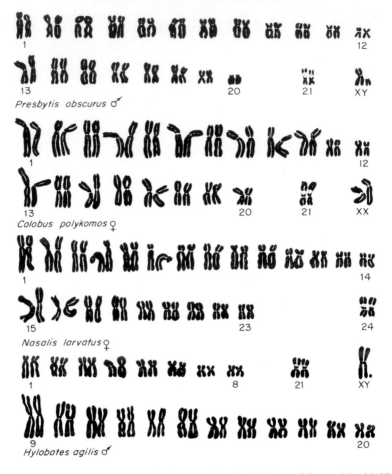

FIG. 11.5 Karyotypes of (a) *Presbytis obscurus* (♂); (b) *Colobus polykomos* (♀); (c) *Nasalis larvatus* (♀); and (d) *Hylobates agilis* (♂).

These data and others which will be given later, give rise to criticisms of the traditional taxonomic organization which places the Colobinae among the Cercopithecidae, and the Hylobatidae among the Hominoidea.

5. THE CHROMOSOMES OF THE ANTHROPOID APES AND THE ORIGIN OF THE HUMAN KARYOTYPE

The number and the morphology of the chromosomes of the anthropoid apes and man are by now known in detail. The number of chromosomes of

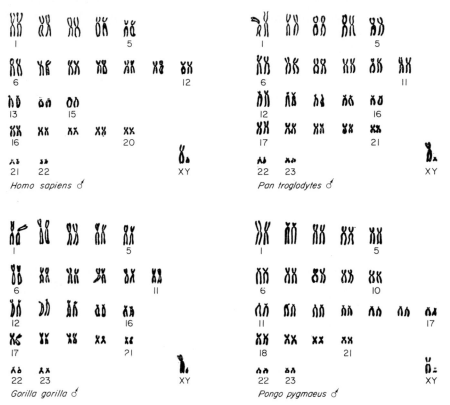

FIG. 11.6 Karyotype of *Homo sapiens*, *Pant troglodytes*, *Gorilla gorilla*, *Pongo pygmaeus*.

the three species of living anthropoids is 48. Man has 46 chromosomes. (Fig. 11.6).

The homologies between these structures in apes and in man can be tested in several ways. A strict morphological comparison gives an approximate idea of the eventual gross variations (i.e. different number of metacentric, submetacentric or acrocentric chromosomes and their different sizes, etc.). It would perhaps be better to make comparisons of chromosomal indices and relative length as this would tend to eliminate the effect of spiralization, or to compare the patterns of incorporation of tritiated thymidine. In the first comparison the results would be limited by differences existing between plates prepared under different conditions, and by errors introduced during the calculations of indices and relative lengths. In the second comparison, morphologically similar chromosomes of Man and Apes may differ in the patterns of thymidine labelling due to physiological asynchrony of their replication. A direct comparison could, therefore lead to some confusion.

Our present approach in comparing karyotypes is to make a direct comparison of the chromosomes obtained from plates with identical total length (some degree of spiralization) in mixed cultures of individuals of two species. The use of mixed cultures eliminates the unavoidable differences existing between slides, especially if they are also prepared in the same laboratory.

A prerequisite for a direct comparison of karyotypes of different species is to know the absolute identity of the quantity of chromosomal material belonging to each species. The exact identity of the amount of chromosomal material can be obtained by measuring the nuclear DNA content (Manfredi-Romanini, 1968) and measuring the total length of karyotypes (Chiarelli, 1972). Comparisons of this kind have up to now been developed only for the chimpanzee and man.

Several results from mixed culture preparations have been identified as being either chimpanzee or man by counting the chromosomes and measuring them. Plates with the same total chromosomal length were karyotyped afterwards according to the Denver system of classification.

To facilitate the comparison of the human karyotype with those of the chimpanzee, a haploid set of human chromosomes is placed in columns as demonstrated in Fig. 11.7. The chromosomes of a metrically identical plate from chimpanzee were then matched with those in the column. Each column corresponds to a group in the Denver system for man. Within the column the chromosomes on the left are man and the chromosomes on the right, chimpanzee.

This karyotype comparison shows clearly that some chromosomes of the chimpanzee are morphologically and dimensionally very similar or almost identical to the corresponding chromosomes in man. Other chromosomes show minor differences such as in the centromeric position or in the length of the arm; they can be considered a homologue to the human ones for at least part of their length. Other chimpanzee chromosomes do not show any real correspondence with the human ones; however, they have been grouped aligned with certain human chromosomes because of a general similarity in size.

There is, however, a chromosome in the human karyotype. Chromosome no. 1 which does not seem to show any counterpart in the karyotype of chimpanzee. On the other hand, the chimpanzee karyotype has two acrocentric chromosome pairs (numbers 12 and 16) which do not show similarities to any human chromosome.

It seems, therefore, quite possible to hypothesize that chromosome no. 1 of man resulted from a centric fusion mechanism between two acrocentric

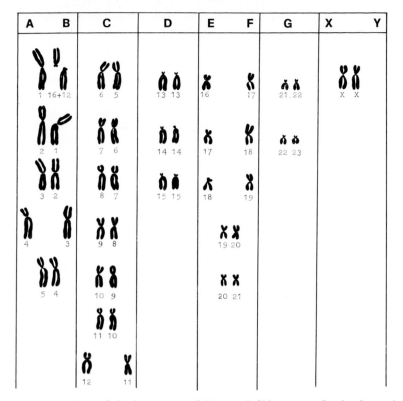

A	B	C		D	E	F	G	X	Y
1 16+12		6 5		13 13	16	17	21 22	X X	
2 1		7 6		14 14	17	18	22 23		
3 2		8 7		15 15	18	19			
4 3		9 8			19 20				
5 4		10 9			20 21				
		11 10							
		12 11							

FIG. 11.7 Comparison of the karyotypes of Man and Chimpanzee. In the first column a haploid set of the chromosomes of Man, and in the second, of the Chimpanzee.

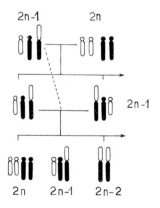

2n-1   2n

2n-1

2n   2n-1   2n-2

FIG. 11.8 Schematic representation showing the way in which a chromosome mutation is established in a population (in this case centric fusion results in the reduction of the number of chromosomes).

chromosomes in an early ancestor of man who was closely related to the chimpanzee. This centric fusion might account for a reduction of the chromosome number from 48 to 46 in one of man's ancestors. As we hypothesized early in 1961, this fusion created an important divergence between the karyotypes of anthropoid apes and man. (Fig. 11.8).

The other differences between chimpanzee and man, which are not of minor importance, could be due to a mechanism of translocation or inversion which certainly took place during the gradual phyletic diversification between the two species. A similar study with two other apes (especially *Gorilla*) will probably tell us which of these variations belong to chimpanzee and which are part of the human line.

An example of a chromosomal rearrangement probably due to a paracentric inversion is that which led to the differentiation of the two species of chimpanzee (*Pan troglodytes* and *Pan paniscus*). These two species and geographical races seem to differ from each other for at least one pair of chromosomes, the twelfth, which is slightly more metacentric in *Pan paniscus* than in *Pan troglodytes*.

However, to return to the transformation from 48 to 46 chromosomes, can we speculate when and how could such a transformation have taken place and differentiated the human karyotype so markedly from that of the anthropoid apes ? The frequency of such mutations is not very rare. Studies of the karyotype of present-day human populations have uncovered, as we have noted before, cases of fusions that occurred long ago between two acrocentric chromosomes with the formation of a karyotype of 45 chromosomes. If two of these individuals were to mate, offspring having 44 chromosomes would be produced, and would probably be perfectly normal. In an actual population it is unlikely that this would happen. Individuals having a karyotype of 45 chromosomes produced by centric fusion are fairly rare, and a mutation of this sort can easily be lost.

In order to understand how this could have occurred in an ancestral prehominid it is first of all necessary to consider the demographic dimensions of the reproductive group which constituted the populations at that time. They could not have exceeded twenty-odd individuals.

They were family clans, very often dominated by a single male who enjoyed absolute dominion over the females of the group. It must have happened fairly frequently that some of the males mated with their own daughters. Indeed it appears that the "leadership" of a group of chimpanzees lasts more than ten years, and the females reach sexual maturity at eight years.

A male with such a mutation; that is, with a karyotype of 47 chromosomes, therefore had two possibilities of establishing this new karyotype, and completing the reduction of the number of chromosomes to 46. One possibility would have been that mating with the various females of the group, produced offspring half of whom would have had 47 chromosomes in the resulting generation and the reduction could have been completed; the other possibility would have been mating with his daughters by a sort of backcross. In this case the process of reduction of the karyotype would have been even more rapid and could have taken only two generations, at the most, therefore, thirty years, considering the shorter time necessary for the acquisition of sexual maturity in our monkey-like ancestors.

The chances of establishing such a mutation obviously are conditional not only on the existence of small reproductive communities but also on the intrinsic power of survival of this mutation.

It is clear that there is little that can be said on this point, yet the massing of a larger number of genes in the same chromosome may have represented some selective advantage at a certain moment in the evolution of the Hominids.

When did this event occur ? As yet it is impossible to say, but as knowledge of the karyotypes of the anthropoid apes increases, a means of attempting a calculation based on the frequencies of chromosomal mutations in natural populations and on the length of generations may be discovered.

6. An Attempt to Revise the Classification of the Catarrhine Monkeys and to Interpret their Phylogenesis on the Basis of Karyological Data

The first data brought to light on the number and morphology of the chromosomes of the catarrhine monkeys, which are in part outlined here, have led to an attempt to revise the taxonomic groupings, especially at the supergeneric level. Undoubtedly, it is at this level that data of this type are most useful.

In Table 11.5 the most significant information with regard to this objective is recapitulated. The quantitative data are tabulated on the left, the qualitative on the right. The most unpredictable quantitative data are the numbers of chromosomes, which vary from 42 to 72.

All the species of the genera *Macaca, Cynopithecus, Papio, Theropithecus, Cercocebus* have the same number of chromosomes ($2n=42$), the same total length of genome, and although the data have not been completely worked out, the same DNA content per nucleus. They have autosomes, marked chromosomes and Y chromosomes of the same type. Moreover, they crossbreed, producing hybrids that are often fertile. It is evident that they belong to a group of genera closely united systematically.

The different species of the genus *Cercopithecus*, on the contrary, have a

TABLE 11.5. Karyological and hybridological data available for a taxonomic revision of the Old World Primates to a supergeneric level.

Quantitative data			Qualitative data					Hybridological Data
Total length of chrom. in M	2n	Genus	Type of chromosome M.	S.	A.	Marked chrom.	Y chrom.	
92 ± 12	42	*Macaca*	6	13	—	A	b	
90 ± 10	42	*Cynopithecus*	6	13	—	A	b	
88 ± 10	42	*Papio*	6	13	—	A	b	
89 ± 10	42	*Theropithecus*	6	13	—	A	b	
85 ± 10	42	*Cercocebus*	6	13	—	A	b	
94–125 ± 10	54–72	*Cercopithecus*	6–9	12–17	6–10	C	a,b,c	(?)
94 ± 10	54	*Erythrocebus*	6	12	7	C	b	—
—	—	*Pygathrix*	—			—	—	—
—	—	*Rhinopithecus*	—			—	—	—
—	—	*Simias*	—			—	—	—
—	48	*Nasalis*	*8	15	—	A	—	—
83 ± 10	44	*Presbytis*	7	12	1	B	a	—
93 ± 10	44	*Colobus*	7	13	—	B	—	—
85 ± 10	44	*Hylobates*	11	9	—	B	c	—
—	50	*Symphalangus*	12	11	1	—(?)	c	—
83 ± 10	48	*Pongo*	—	12	11	—	a	—
94 ± 10	48	*Pan*	5	10	8	—	a	—
98 ± 10	48	*Gorilla*	5	10	8	—	a	—
93 ± 10	46	*Homo*	4	13	5	—	a	—

The letters A, B, and C indicate the three different types of marked chromosomes: (A) with an arm opposite a large achromatic region; (B) with an arm opposite a smaller achromatic region; (C) with an arm opposite an achromatic region which is nearly invisible. The letters a, b, c indicate Y chromosomes differing in size and position of centromere: (a) large and acrocentric or subacrocentric; (b) metacentric; and (c) tiny and dot-like.

variable number of chromosomes ranging between 54 and 72. *Erythrocebus* has 54 chromosomes. The total length of the karyotype is also variable and increases in direct relation to the number of chromosomes. Three types of chromosomes are present (metacentric, submetacentric, acrocentric); their relative numbers increase more or less regularly with the increase in the total number of chromosomes. The marked chromosome has a very short or barely existent non-marked branch. The size of the Y chromosome is variable.

Unfortunately, the karyotypes of the genera *Pygathrix, Rhinopithecus* and *Simias* are as yet unknown.

Nasalis larvatus has 48 chromosomes. These appear to be very similar morphologically to those of *Presbytis* and *Colobus*; the marked chromosome also resembles that of these genera.

Presbytis and *Colobus* both have the same number ($2n=44$) and do not display significant differences in the total length of their genome. The chromosomes of these two genera appear to be fairly similar morphologically. The number of chromosomes of the gibbons (*Hylobates*) is the same as that of *Presbytis* and *Colobus* ($2n=44$). The total length of the genome is the same. In the general morphology of the chromosomes the metacentric ones are prevalent. The marked chromosome is almost identical with that of *Presbytis* and *Colobus*. The Y chromosome is the smallest among those observed to date in the Old World monkeys.

The karyotype of *Symphalangus* is very different from that of *Hylobates*. The number of chromosomes is $2n=50$, and they differ somewhat. The marked chromosome is missing, even though an acrocentric chromosome with an achromatic terminal region suggests a remnant of the marked region. Consequently, *Symphalangus*, from the karyological point of view, appears fairly different from *Hylobates*.

The true anthropoids (*Pongo, Pan, Gorilla*) have the same chromosome number ($2n=48$) and the same length of genome. A chromosome marked by an achromatic region of the type found in the other Old World primates is missing from the anthropoid apes.

The karyotypes of the gorilla and chimpanzee are very similar to each other. The karyotype of the orangutan differs from those of the gorilla and chimpanzee in various morphological particulars.

Man differs from the anthropoid apes in the morphology of some chromosomes.

The anthropoid ape whose karyotype is most similar to the human one is the chimpanzee.

Notwithstanding the lacunae which still exist, some preliminary taxonomic conclusions can be drawn from these data.

The genera *Macaca, Papio, Theropithecus,* and *Cercocebus* must be separated from the species belonging to the genus *Cercopithecus* and placed in a different subfamily to which the name Papinae could be assigned, leaving the name Cercopithecinae to the species belonging to the genus *Cercopithecus* only.

The genus *Symphalangus* must be separated from the various species of *Hylobates*.

Both these genera must then be removed from the superfamily Hominoidea: they could be included in the superfamily Cercopithecoidea and constitute a family in themselves.

A detailed discussion of the group of Colobidae is still not possible at the moment, yet they appear to be a fairly homogeneous group.

The superfamily Hominoidea should be restricted to the anthropoid apes (*Pongo*, *Gorilla*, *Pan*) and to man. Among these man is closely distinguished by the number of chromosomes and must be classified in the family Hominidae; while the true apes will constitute the subfamily Pongidae.

In the subfamily Pongidae the karyotype of the orang can be distinguished clearly from that of the gorilla and the chimpanzee. This difference could be taken into consideration at a supergeneric level, to distinguish the orang-utan from the African apes.

The morphological variations in the chromosomes of the different species may be the result of different mechanisms, the most common of which are inversion and translocation.

Nevertheless, from a phyletic point of view the variation in the number of chromosomes is of greater interest. The mechanisms which lead to these numerical variations, as we have said, are centric fusions, centric fission, polyploidy and polysomy.

Which of these mechanisms is responsible for the numerical variations in this group of species? What was the original chromosome number of the ancestor common to all the Old World primates, accepting that we have had a single common ancestor? The chromosome number in the somatic cells in the living species of Old World primates varies from 42 to 72, but the greater part of this variability (from 54 to 72) can be attributed to the diverse species of the genus *Cercopithecus*. Extremely discordant opinions exist concerning the mechanisms which have led to such a wide variation among the different species of a single genus, which moreover, are very homogeneous from an anatomical-physiological standpoint. It is therefore impossible to form an hypothesis as to when this group was separated from the other species of Old World primates, or to which group of *Cercopithecus* they are most closely related.

In the other species of Old World primates the number of chromosomes varies from 42 to 50. The original number of chromosomes in the possible ancestor of the Old World primates must probably be sought within the limits of these eight pairs. The fact that the groups of taxonomically diverse and phylogenetically ancient species such as *Presbytis*, *Colobus*, and *Hylobates* all have 44 chromosomes suggests the hypothesis that this was the number of chromosomes of the species of primate that, in the middle Eocene about fifty million years ago, was the original ancestor Old World primate.

A centric fusion between two pairs of acrocentric chromosomes or a double translocation could have been responsible for the reduction from 44 to 42. For an interpretation of the chromosome number 48 in *Nasalis* and 50 in *Symphalangus* recourse must be had to mechanisms of centric fusion or of polysomy. Deeper research into the quantity of DNA in these species could contribute towards clarifying this problem.

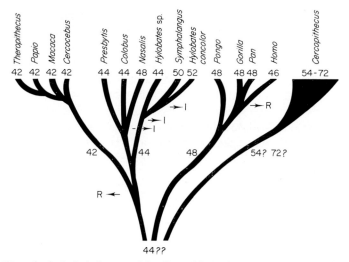

Fig. 11.9 Hypothetical phyletic tree of the Catarrhine primates based on karyological data. The numbers indicate the diploid (2*n*) number of chromosomes. R indicates the reduction mechanism of chromosome fusion; I indicates that polysomy exists for some of the chromosomes.

Probably the number and morphology of the chromosomes of the anthropoid apes are not directly connected with the primate forms mentioned before.

The general morphology of the chromosomes is very different, and if there was a common ancestor he must be sought in the very remote past. The origin of the chromosome number of 46 for man has been discussed in a preceding section.

From the karyological point of view, therefore, we can distinguish three distinct lines of evolution in the groups of catarrhine monkeys as schematized in Fig. 11.9: that concerning the group of species of the genus *Cercopithecus*; that of the various genera *Macaca, Papio. Theropithecus, Cercocebus, Presbytis, Colobus, Symphalangus, Hylobates,* and *Nasalis*; and that of the three anthropoid apes and man. The connections between these three distinct lines of evolution as yet seem indistinct and controversial.

7. Chromosome Banding and DNA Hybridization

There are several relatively new techniques for studying chromosomes that promise to elucidate many problems in comparative and evolutionary karyology. These techniques give information about the structural and physiological differences along the length of chromosomes, and their results are so consistent that they can be used to compare homologous chromosomes in the same or different metaphase plates both among and between species.

One technique employs several dyes; i.e. quinacrine dihydrochloride (Q) and quinacrine mustard (QM) to produce fluorescent banding patterns along the chromosome. Giemsa stain can also be used selectively. Pretreatment with trypsin before staining seems to allow the stains to distinguish the sections of the chromosome more easily.

The best method found thus far is DNA hybridization *in vitro*, where DNA strands from two species are dissociated into single strands and then recombined into hybrid double strands. This can often demonstrate homologies between different species. Martin and Hoyer (1967) have had some interesting results which open up unexpected avenues of research on the chromosomes of primate species.

GENERAL REFERENCES

Bender, M. A. and Chu, E. H. Y. (1963). The Chromosomes of Primates. *In* "Evolutionary and Genetic Biology of Primates" (J. Buettner-Janusch, ed.), Vol. 1, pp. 261–310, Academic Press, New York and London.

Chiarelli, B. (1962). Comparative morphometric analysis of Primate chromosomes. II. The chromosomes of the genera *Macaca, Papio, Theropithecus,* and *Cercocebus. Caryologia* 15, 401–420.

Chiarelli, B. (1962). Comparative morphometric analysis of Primate chromosomes. I. The chromosomes of anthropoid apes and of Man. *Caryologia* 15, 99–121.

Chiarelli, A. B. (1963). Observations on P.T.C.-Tasting and on hybridization in Primates. *In* "Symposia of the Zoological Society of London" 10, 277–279.

Chiarelli, B. (1963). Comparative morphometric analysis of Primate chromosomes. III. The chromosomes of the genera *Hylobates, Colobus,* and *Presbytis. Caryologia* 16, 637–648.

Chiarelli, B. (1968). Caryological and Hybridological Data for the Taxonomy and Phylogeny of the Old World Primates. *In* "Taxonomy and Phylogeny of Old World Primates with References to the Origin of Man", Rosenberg and Sellier, Turin.

Chiarelli, B. (1968). Chromosome polymorphism in the genus *Cercopithecus. Cytologia* 33, 1–16.

Chiarelli, A. B. (1972). Comparative chromosome analysis between Man and Chimpanzee. *J. Hum. Evol.* 1, 389–393.

Egozcue, J. (1968). Primates. *In* "Comparative Mammalian Cytogenetics" Benirschke, ed.), Springer-Verlag, Berlin, Heidelberg, New York.

Manfredi-Romanini, M. G. (1968). Quantitative relative determination of the nuclear DNA in the Old World Primates. *In* "Taxonomy and Phylogeny of Old World Primates with References to the Origin of Man", Rosenberg and Sellier, Turin.

Manfredi-Romanini, M. G. (1972). The nuclear content and area of primate lymphocytes as a cytotaxonomical tool. *J. Hum. Evol.* 1, 23–40.

Martin, M. A. and Hoyer, B. H. (1967). Adenine plus thymidine and guanine plus cytosine enriched fractions of animal DNAs as indicators of polynucleotide homologies. *J. molec. Biol.* 27, 113–129.

$I2$ ═══════

Palaeontological Data as Proof of the Differentiation of Species

1. GENERAL

In the preceding chapters we have tried to enumerate the mechanisms by which the various primate species could have been differentiated from a possible common stock.

However, up to this point we have not provided any concrete proof in support of this thesis. Certainly, we have seen how small variations could arise and establish themselves, or how systematic relationships could be substantiated even between widely separated groups of species, but we have not yet offered proofs of the actual existence of intermediate forms. The order of classification which we have proposed in Chapter 5 is not arbitrary; it is based on significant anatomical, functional and genetic characters which make such a disposition logical. That is, it is based on real similarities between the different groups of primates.

The first set of facts which enables us to determine possible relationships between various organisms is based on comparative morphology. The resemblances proclaim a kinship, a shared derivation, a descent from common ancestors.

The most closely related forms in the natural system also present the greatest similarities. The greater the resemblance the closer the genetic homogeneity.

A complete natural system, that is, one that also comprises fossil forms, corresponds to a genealogical tree; if it were perfect it would give us detailed information on the course followed by evolution. In other words, only with the fossil finds does a taxonomic classification become phyletic. The incontestable proof is, therefore, provided by fossils; that is, by the study of the relics of beings who lived in remote times, whose remains can be found embedded in sedimentary rocks.

Unfortunately, the documentation of these remains is not always complete; on the contrary, it is very often fragmentary, and only for short periods of phylogeny is there any possibility of having a complete series. Very often it is not only single connecting links that are missing, but entire lengths of the chain.

Palaeontology, alone, cannot reveal how evolution came about, but it irrefutably demonstrates that evolution did occur. This is because palaeontology supplies information that the living things of the past were different from those of the present: the further back in time, the more different they were. However, for a correct interpretation of fossil remains great caution must be exercised.

2. Problems and Precautions in the Comparison of Fossil Remains

In the comparison of morphological characters it is first of all necessary to give proper weight to their complexity.

Without doubt much of the dispersion of the hominids and other primates is based on the different valuation of single characters. It is impossible to affirm with certainty that a single structural detail or one measurement can in itself be accepted as proof that a certain fossil fragment belongs to a hominid and not to another primate (see also Chapter 10).

The presence of a space (diastema) between the canine and the incisor in the jaw is not in itself sufficient proof to determine that a fragment of a mandible belongs to the fossil of an anthropoid rather than to a hominid. There are alterations due to the particular form of the canines or to the lateral displacement of the first premolar that will mask the presence or absence of the diastema.

Secondly, the possibility of convergent or parallel evolution of some characters in different groups of animals must be kept in mind. In other words, it may happen that two or more species which display an initial similarity of structure and the same type of adaptation, then undergo a recurrent succession of analogous mutations leading to parallel or sometimes convergent evolution.

A third major difficulty, not, however, intrinsic to the material, consists of the multitude of names (generic and specific) which the various palaeontologists are accustomed to attribute to even very fragmentary finds of primates. The group of Australopithecines is a typical example of this proliferation of the nomenclature. Dart attributed the specific name *Australopithecus africanus* to the first fossil finds. Successive investigators found many other remains of individuals belonging to the same group which they separated into almost as many different genera as species

(*Plesianthropus transvaalensis, Paranthropus robustus, Paranthropus crassidens, Telanthropus capensis,* etc.). More recently another name, *Zinjanthropus boisei,* has been attributed to a find in southeast Africa. Current opinion now holds that there is no sufficient reason for distinguishing these finds at the generic level, and also that the number of species must be reduced to two or three at the most.

Finally, in order to make the comparison of the morphological characters objective, repeated attempts have been made to evaluate the degree of similarity on quantitative bases. These attempts, in themselves valuable, often have not given adequate consideration to the relative importance of the different characters in taxonomy and the difficulty of adequate measurement which many of these finds present. In general, statistical comparison of the total measurements of the indices is extremely useful in establishing degrees of affinity between forms that are certainly related. However, data of this type have been of less practical value when the relationship between the various forms to be compared is more remote.

3. THE EVOLUTIONARY SERIES OF THE EQUINES

The equines will be used in this section as an example of a well-known evolutionary series within the ambit of the mammals.

In the suborder of the Hippomorphs, that of the equines is the most important family, represented today by three groups of species: the horse (*Equus*), the zebra (*Hippotigris*) and the ass (*Asinus*). These still possess a certain idioplasmatic continuity, demonstrated by the possibility of obtaining hybrids from different species, even if most of them are sterile.

Among other characteristics they are united in having only one toe on each foot (because of this they are classified in the order of the Perissodactyla), a rather agile body structure and highly specialized dentition. They are all forms very well adapted to running on grassland (Fig. 12.1). Because of the type of pasture, based principally on the grasses, and of the related type of mastication, their dentition consists of three large incisors, a rudimentary canine followed by a wide diastema, a first premolar and three molars of increasing size, with complicated masticatory ridges, semilunar in form.

The most ancient forms differ radically from those of present-day species, and it is only from the intermediate stages that it is possible to realize that these are related to existing forms.

The phylogenesis of the equines is fairly well known. They originated essentially in North America, whence there was a repeated movement of some species to Eurasia, and from the Eocene onwards also into South America (Fig. 12.2). The most ancient genus known is the *Eohippus* of the

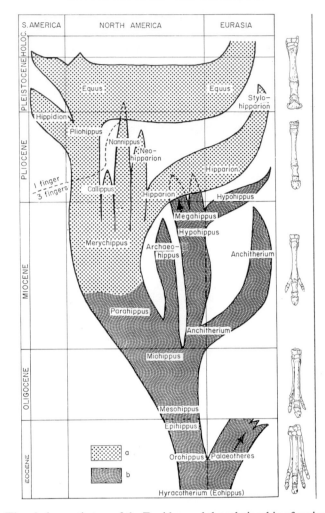

FIG. 12.1 The phylogenetic tree of the Equidae, and the relationship of various forms with changing feeding patterns according to Simpson: (a) Grass feeding; (b) Bush feeding.

Lower Eocene, more or less the size of a hare. It had a complete dentition with the diastema scarcely hinted at, and little or no specialization in the premolars and molars. The foreleg rested on four toes, the hind leg on three; the nail of the middle toe was hoof-like and the animal walked on its toes. This form, which probably lived in the forests browsing on the leaves of shrubs, also diffused into Eurasia where it developed into a group of very interesting forms (*Hyracotherium*), with no relation to the phylum of existing equines. From the *Eohippus* of North America there were derived,

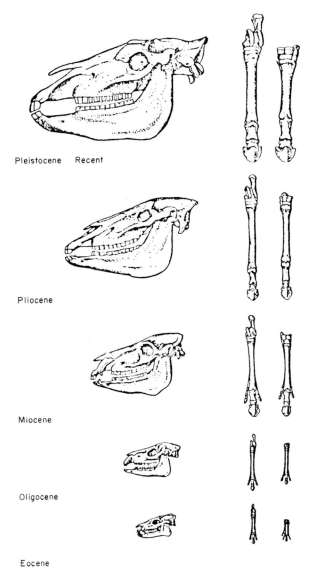

Pleistocene   Recent

Pliocene

Miocene

Oligocene

Eocene

Fig. 12.2 The crania and posterior and anterior limbs in different types of fossil Equidae (on top is the modern horse) (according to Crow, 1959).

via numerous phyletic branches, a great many species (about 350).

Eohippus in America developed into a form of the dimensions of a goat (*Mesohippus*) during the Oligocene; *Mesohippus* had three toes of which only the central one touched the ground. Its teeth were still brachiodonti, but all except the incisors had assumed the aspect of molars. During the Miocene this form developed into a large number of types; the teeth becoming larger in each case.

Among the others one form (*Hipparion*), which was the size of a pony, reached Europe with a new migratory wave at the beginning of the Pliocene and lived there up to the Pleistocene. It was very similar to the existing horse and for a long time was considered one of its precursors. Nevertheless, the ancestor of the present-day horse is the *Pliohippus*, which is also American. *Equus* arrived in Europe following a new stream of migration at the beginning of the Pleistocene. For some unknown reason, probably an epidemic, horses disappeared in America at the end of the Pleistocene and were re-introduced there by man only during the last few centuries.

Various collateral ramifications, with species which had a limited survival, are joined to this principal line of evolution.

In the main line evolutionary development led little by little to a better adaptation for running and to grassland pasture. The most important modifications brought about during evolution were the increase in size, the elongation of the metapodalia, the reduction and disappearance of the lateral toes, the elongation of the muzzle, the transformation of the premolars into molars, the complication of the grinding surfaces of the teeth, and the closing of the posterior margin of the orbit, with consequent separation of the orbital cavity from the temporal fossa, because of the necessity for a better attachment of the masticatory muscles. In the Oligocene *Mesohippus* and *Miohippus* still rested three toes on the ground and probably still browsed on leaves of shrubs. Life on the plains and a grass diet were initiated during the Miocene with *Parahippus* in which the second and fourth toes still had a little hoof, but the weight of the body was borne entirely by the third.

The present-day horse is separated from the tiny *Eohippus* by an interval of time of about 50 million years. *Eohippus* represents a miniature copy of the *Equus*. The differences between these two forms were annulled by a multitude of intermediate types developed during this lapse of time.

In the same way as in the case of the horse, the various stages in the evolutionary history of the elephant, of many ungulates and other animals of the plains have been reconstructed. The history of the primates, as we shall see, is more difficult. In fact, because of their life in the forest, they have left rare and usually badly preserved fossil remains. For this reason every item of information, no matter how fragmentary, must be treasured.

GENERAL REFERENCES

Clark, W. E. Le Gros (1964). "The Fossil Evidence for Human Evolution", The University of Chicago Press, Chicago.

Padoa, E. (1963). "Manuale di Anatomia comparata dei Vertebrati", Feltrinelli, Milano.

Simpson, G. G. (1965). "The Geography of Evolution", Churchman, Philadelphia.

Trevisan, L. and Tongiorgi, E. (1958). "La Terra" UTET, Turin.

13

Time in the Evolution of the Primates and of Man: Methods of Measuring it and its Subdivisions

1. STRATIGRAPHIC SUBDIVISIONS

It is far easier for us to grasp the abstract concept of time when we think of it in familiar intervals—days, years, centuries. Once we step outside these accustomed limits, it becomes necessary to pause and, very often, derive our impressions from a reduced time scale. Thus, for example if we alter the period of time for the appearance of the first vertebrates to two months ago, with man arriving on the scene during the last hour it is clearer than if we state that the first vertebrates appeared on earth six hundred million years ago and man only five to six hundred thousand years ago.

How is it possible to reconstruct the pattern of succession of the diverse forms of life on earth? The first evidence is furnished by the succession of parallel strata in sedimentary rocks. In fact, in a regular series of rocks of sedimentary origin, if undisturbed, the relative ages of the strata are calculated according to a clear and logical principle, which is called the law of superposition. The lowest stratum must be the most ancient, that nearest the top the most recent. The law of superposition is the basis for the calculation of geological age and also for the theory of the evolution of living things, since it is possible to demonstrate the succession of events and the development of life throughout geological time by referring to it.

Still, the succession of the strata is not always regular. Various natural phenomena may have altered the order, overturning the entire series of strata or at least producing disturbances. In such cases a most effective assistance is provided by the fossil remains of living things, which are called guide-fossils. Each group of strata, characteristically, contains a typical group of fossils, which cannot be found in identical form in any other formation, either above or below.

In the identification of geological structures the guide-fossils have the same importance as have coins, potsherds and artifacts in defining the sequence of periods from the remains of a civilization which flourished in a particular area in the past.

All traces of ancient life in the rocks are fossils: animals, plants, footprints, impressions of bodies, shells, skeletons. In general soft parts decompose and disappear; only the hard parts remain, nor are they always in good condition.

Just as coins of a particular type found in distant regions indicate that the strata that contain them must have received the same cultural influences contemporaneously, so the finding of fossils of the same type in two far distant strata indicates a chronological correspondence between them.

By means of the law of superposition and the examination of the anomalies and of the fossil associations it is possible to construct a geological series 150 km thick, which comprises a period of more than 1000 million years.

This geological series, which represents the story of life on earth, has been subdivided into volumes; sections; chapters, which to geologists are the eras, periods and so on. We are interested in the last two eras: the Tertiary and Quaternary. The Tertiary, or Cenozoic in which the development of the mammals took place, lasted about 80,000,000 years and is subdivided into five periods: Palaeocene, Eocene, Oligocene, Miocene and Pliocene; the Quaternary in which we live, began about 1,000,000 years ago and is subdivided into the Pleistocene and Recent.

2. The Search for an Absolute Chronology

The thickness of sedimentary deposits seems an obvious choice as an indicator of duration in any attempt at an absolute chronological estimate. The estimate of time would be automatic if the rate of sedimentation had always and everywhere been constant, and if we knew what thickness of sediment accumulates in any unit of time. Unfortunately the speed of sedimentation varies between limits that are too wide to permit a calculation of this type to give reliable results. Therefore, this method will not serve to determine the absolute age of sediments.

In some particular cases the duration of the formation of a stratum is re-vealed by the dimensions of the granules, and the regularity of these varia-tions repeated many times can be attributed to a constant rhythm of known duration. When the rhythm of sedimentation has an annual oscillation, such strata are called "varves". The study of rhythms of longer duration is more difficult.

3. The Varve Method

This method based on the principles of sedimentation yields absolute results. It is of great interest because it is particularly applicable to the last glaciations.

The word "varve" is derived from the Swedish "verving" which means cycle, and has entered into international terminology with the significance of "sedimentary stratum formed in the course of a year". Each varve is made up of a lower part of very fine sand and an upper one more clayey and darker in colour. A thin veil of blackish clay clearly separates one varve from the next.

Deposits of this type are discontinuous. They are found most extensively in Sweden and Finland; countries in which the retreat of the glacial ice-cap has taken place during the last 20,000 years more or less. The most ancient deposits are found on the southern shores of the Baltic, the more recent move progressively towards the north.

The study of varves consists in establishing correlations in time between a series of varves in one place and other series in places both closer and farther away. The absolute thickness of a varve has little significance: of more in-terest are the relative differences between many contiguous varves. The clearest part of the larger varves is most often that which represents the summer sedimentation. Consequently comparatively thick varves represent hot or especially long summers, while thin varves denote short, cool sum-mers. The thickness, therefore, varies with a rhythm which depends on the meteorological and climatic differences of the various years. These data can be depicted diagrammatically. In this way a breakdown is obtained which represents an interesting document of "fossil meteorology"; that is, the natural record of as many consecutive summers as there are varves reported (Fig. 13.1).

Correlations are established by comparing diagrams of different places. Sometimes the varves are not formed of glacial mud but of the successive seasonal laying down of shells of diatoms. Because of the particular system of reproduction of the diatoms the generations which succeed each other in the various seasons have shells of different sizes. Thus it is possible to study the size of the shells statistically and identify the spring, summer and

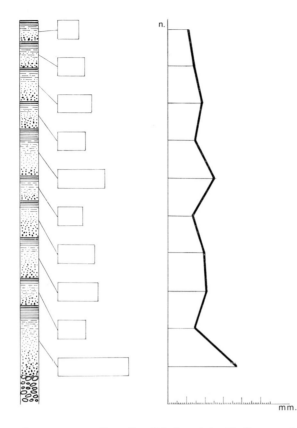

FIG. 13.1 Annual varve pattern (from Swedish deposits) with diagram obtained from this on the right. The average thickness of each varve is 3–4 mm.

autumn generations in each single varve. It is evident that the thickness of annual deposits is in some way connected with climatic variations in the case of lacustrine varves as well.

Complexes of strata exist in Italy which can in all probability be considered annual varves. During the middle part of the Quaternary, when the volcanoes of Mt. Amiata and of Bolsena were active, there were a very large number of lakes in these regions, in which sedimentation developed under very special environmental conditions. The sediments were formed exclusively of the calcareous shells of diatoms, which found an exceptionally favourable environment for multiplication in those waters. The strata are one to two centimetres thick, and are composed of two layers: a white one made up more or less exclusively of shells of diatoms, and a brown one made up of shells of diatoms mixed with organic substances. Sometimes in the dark layer it is possible to recognize the impression of leaves, which probably represents the autumnal deposit.

4. Method Based on the Annual Growth of Plants

The transverse section of a plant which has an annual growth period yields information in the same way as varves regarding the succession of years within the limits of a determinate period, and on its climate. As with varves, it is possible, within certain limits, to correlate annual cycles in different plants and to reconstruct the sequence of years even over long periods of time (Fig. 13.2).

5. The Fluorine Method

Chronology based on varves and on annual growth of plants is limited in space and in time. Only occasionally is it possible by means of these systems to reconstruct a long and continuous time sequence.

The palaeontologist needs methods which have a general value and which permit the absolute dating of single objects. One such possible method is provided by the study of the accumulation of fluorine in the bones of vertebrates.

If there has not been interference with the earth, the bones of any creature remain buried in the same soil on which it lay at the moment of death. There they are in continuous contact with solutions in which fluorine is

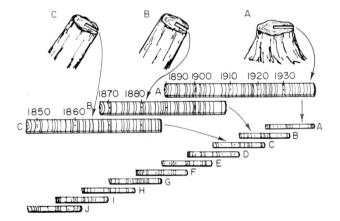

Fig. 13.2 Diagram illustrating the chronology based on the annual growth rings of trees (A) living tree; (B) beam from a recent house; (C) beam obtained from an ancient building. By superimposing the growth rings of the trees it is possible to date backwards into prehistoric times (D–J) (drawing prepared by F. Fedele).

always present, even if in minute quantities. The mechanism of penetration of fluorine into the bones involves a process of the substitution of the OH group of the phosphates of the bones [hydroxyapatite, $3Ca_3 (PO_4)_2$ Ca $(OH)_2$] with the consequent formation of an almost insoluble compound [(fluorapatite) $3Ca_3 (PO_4)_2$ Ca F_2].

It is fairly obvious that in this case the relationship between time and the quantity of fluorine present in fossil bones cannot be kept constant. The number of OH groups susceptible to substitution in fact decreases little by little as the substitution proceeds. Comparatively rapid at the beginning, the phenomenon becomes steadily slower as time passes, according to laws which must be determined region by region, depending partly on the quantity of fluorine in the underground water. When the quantity of fluorine in the surrounding water is relatively low and the region subjected to high evaporation, the fluorine content of the fossil bones reaches saturation point in about half a million years.

Consequently samples can be distinguished from one another within these time limits on the basis of their fluorine content as expressing their different ages. Under other conditions the extent of application of the method is lowered to less than 20,000 years.

The surest method is first to determine the fluorine content of various specimens from a region, some of which have already been dated by other means, and then to decide the age of the unknown samples by comparison. Even within the limits imposed by these conditions the method may be applied in many cases, and is especially useful when there are doubts as to the provenance of the specimen. An example of this was the discovery of that celebrated scientific fraud, the Piltdown cranium (*Eanthropus dawsoni*), which was made possible by the use of the fluorine method.

6. THE RADIOCARBON METHOD

Ordinary carbon (^{12}C) constitutes 98·9% of all terrestrial carbon; the remaining 1·1% is made up of various isotopes of which ^{14}C is the most frequent in nature. In nature ^{14}C is formed by the collision of neutrons with atoms of nitrogen.

Artificially it can be obtained in large quantities from a uranium pile by the action of slow neutrons on nitrogen,

$$^{14}N + n^1 = {}^{14}C + H,$$

by the same process by which it is produced naturally. In fact, in the upper strata of the atmosphere the neutrons produced by the collision of cosmic rays with the elements of the atmosphere make ^{14}C out of nitrogen.

The ^{14}C thus obtained reacts with atmospheric oxygen giving rise to radioactive CO_2 which enters into the structure of all living material.

In 1949 Libby propounded the hypothesis of a radioactive equilibrium of ^{14}C, according to which the percentage of ^{14}C disintegrated must be equal to the percentage of ^{14}C produced in the atmosphere. From this hypothesis it was deduced that when life ceases, the acquisition of ^{14}C within plants and animals also ceases, while the disintegration of ^{14}C into ^{14}N continues. Thus the concentration of ^{14}C decreases proportionally with time. Based on the validity of these hypotheses the half-life of ^{14}C was measured.

It has been measured various times by many methods, and has been calculated to be between 5580 and 5513. The formula which is used to determine the age of a material is the following:

$$I = I_0 \, e - \frac{t}{\tau}$$

where:

> I is the value of the radioactivity of the material which is being examined;
> I_0 is the specific activity (per gram and per minute) of the material (living carbon) measured in inorganic substances;
> t is the time;
> τ is the global half-life, which is a constant equal to 8600 (constant of transmutation).

By means of this formula it has been possible to construct the curve of decay of [14]C (Fig. 13.3).

On the same graph the values determined for specimens of known age are reported, showing a perfect agreement with the theoretical dates.

The precision of the method is inversely proportional to the antiquity of the find and depends on various technical factors. Its limit of validity, according to Libby, is about 20,000–25,000 years; according to Suess about 32,000–35,000 years; according to others it may reach higher values (50,000–70,000 years).

7. Other Physical Methods

The radiocarbon method has a maximum validity, as we have seen, of not more than 70,000 years. As a method its chief value is, therefore, archaeological. We are interested in methods which might make datings possible over even millions of years.

Like ^{14}C, all radioactive elements can be dated. As long as the relationship between a radioactive element and its disintegration product is known,

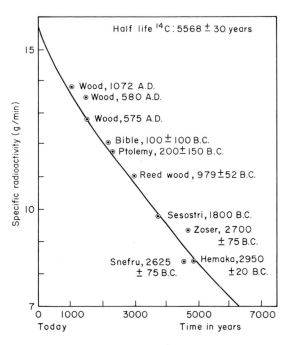

FIG. 13.3 Graph showing the curve of decay of ^{14}C and the age determined for samples which had already been dated using archaeological or historical data.

it is possible to calculate the epoch in which the rock containing it was formed.

Let us suppose, for example, that we have a mineral which at the moment of its formation contained a single radioactive element, thorium 232, with no trace of lead. Thorium, by emitting α particles, gives rise to lead 208, and consequently the mineral, after a certain length of time, becomes richer in lead. Since the amount of lead present depends on the law governing the radioactive decomposition of thorium ($4 \cdot 95 \times 10-11$ years) it is easy to calculate the age of the rock. The same argument holds for rocks containing radioactive uranium (^{234}U, ^{235}U, ^{238}U).

With these methods, numerous age determinations have been obtained, covering the interval from the origin of the Earth to the first phases of the Tertiary (50 million years ago).

However, the precision of the methods is inversely proportional to the proximity to the present era, and, moreover, they do not yield accurate dates for materials from the Tertiary and Quaternary. These latter periods are of the greatest interest as regards the evolution of the primates.

A method which appears to give results of particular interest in this respect is that of potassium-argon. Potassium 40 decomposes in two ways; either giving rise to argon 40 which can be measured and so makes possible an age calculation, or disintegrating into calcium 40, with which dating is possible only if the original mineral contained none of this element at all.

Other methods are based on the modification brought about in a mineral by the bombardment of α particles originating in traces of radioactive elements already contained in the specimen.

It is possible, in this way, to determine the defects in the crystal lattice by the emission of luminous energy when the sample is heated (thermoluminescence). The energy in the form of light seems to depend on the existence of the defects; consequently, if the content of radioactive elements is known the time during which they have acted can be estimated.

8. The Remains of the Products of Human Culture as a Means of Studying the Chronological Sequence of Human Populations

The aim of this and the preceding paragraphs is to enumerate the means which may be used to date, either relatively or absolutely, all fossil finds. Other than skeletal remains, the products of human activity are available, and in remarkable quantities. It is possible to use this material to reconstruct the history of ancient populations, even if the record is very incomplete at present.

The oldest pre-human populations almost certainly used sticks and bones as objects of defence and offence. Even some of the present anthropoid apes use sticks for various purposes, and probably the Australopithecines used femurs and other fragments of bone to kill the animals they ate. Still, these objects have not survived, and their use is only supposition. The products of use which have been most likely to be preserved until our time are worked stones. Certainly the first man, like many still existing and like the apes, used single pebbles as weapons of defence and offence. However, no marks are visible to indicate that they were so used by our ancient ancestors. It is only when we find worked stones that we can assume that they are the fruits of human labour.

It is doubtful that activity of this type was carried on by men of the Tertiary, even if some traces do exist (Ipswich, England). With regard to Pleistocene finds (Quaternary) on the contrary, no doubt can exist about the human origin of certain flaked stones.

The instruments have a clearly defined aspect and with them are found remains of the animals that man hunted and vestiges of the fires that he lit. Human cultures were not uniform throughout all of prehistory; rather, depending on the period at which they arose and the place in which they

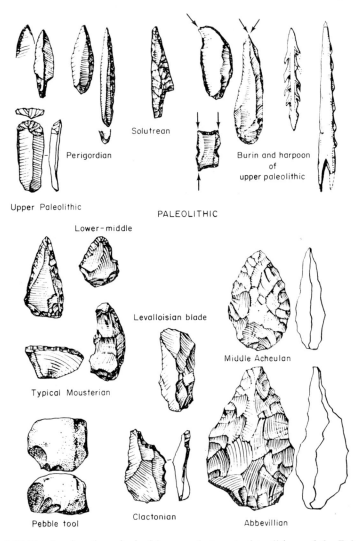

FIG. 13.4 Table showing the principal bone and stone tool traditions of the Palaeolothic in Europe (drawing prepared by F. Fedele).

developed, they display very distinct variations. With the passage of time human industries, from generally crude beginnings, tended to become steadily more perfect and differentiated. However, this trend toward perfection never, or only occasionally, developed in one area. More often the succession was aided by true cultural waves which, arriving from other regions not easy to identify, were gradually superimposed on the *facies* of the pre-existing cultures, submerging them and taking their places. As time passed, the tendency was for each *facies* to endure for a shorter period. The most recent are very brief compared to the ancient ones.

From the study of prehistoric human industries and of the fauna which accompanied them through the different cultural phases, we can, schematically, distinguish a more ancient phase when people were exclusively hunters (Palaeolithic), followed by a later phase of herdsmen and farmers (Neolithic). During all of the Pleistocene man lived by hunting and gathering fruit. He knew nothing of agriculture nor of the domestication of animals. His existence was tied to the enormous herds of animals (bison, mammoth, reindeer) which co-existed with him, and he was forced to follow their migrations, principally caused by changes in climate and vegetation. The sequence of different cultural *facies* in our regions is probably due to variations in climatic conditions.

The activity of hunting stimulated during the Pleistocene the manifestation and development of most of man's higher faculties and the evolution of material culture.

The artifacts of this period consist of variously worked stones (Fig. 13.4). Precisely for this reason it is called the Palaeolithic or period of flaked stone. The Palaeolithic draws to a close at the end of the Pleistocene.

The present geological epoch, the Holocene, began with a new period called the Neolithic. With the coming of the Neolithic we find a radical change in the economy of human societies.

The new populations which reached our regions and which were also slightly different somatically, overcame and eliminated the palaeolithic populations, bringing new cultural elements which soon spread everywhere. Domestication and rearing of animals developed in this period, the construction of huts and villages on piles began, the use of ceramics and weaving were established, leading to the age of metals.

TABLE 13.1. The Cenozoic era in millions of years

Periods	Epochs	Duration each	Began x millions of years ago
Quaternary	Pleistocene and Recent	1	1
Tertiary	Pliocene	11	12
	Miocene	13·7	25·7
	Oligocene	8·3	34
	Eocene	21	55
	Paleocene	23	78
Mesozoic		12∼0	20∼0
Palaeozoic		31∼0	50∼0
Archaeozoic		–	–

9. The Time-scale in the Evolution of the Primates

Finds of primate fossils cover the entire Cenozoic era. The duration of this period is estimated at about 70–80 million years. It is subdivided into periods and epochs, as schematized in Table 13.1.

The subdivisions of the Quaternary have been reviewed in detail in the preceding paragraph and will be discussed again in the next chapter.

General References

Brothwell, D. and Higgs, E. (eds) (1963). "Science in Archaeology", Thames and Hudson, London.

Graziosi, P. (1967). Le civiltà preistoriche. In "Razze e Popoli della Terra" (R. Biasutti, ed.), Vol. I, 134–185, UTET, Turin.

Stallings, W. S., Jr. (1949). "Dating Prehistoric Ruins by Tree Rings", Laboratory of Tree-Ring Research, Tucson, Arizona.

Trevisan, L. and Tongiorgi, E. (1958). "La Terra", UTET, Turin, Italy.

14

Climate during the Period of the Evolution of the Primates and the Appearance of Man

1. Importance of Climate and Climatic Factors

Climate is an important factor in the evolution of a species. Sudden variations of climate, for example causing excesses of temperature or of humidity, may lead to the extinction of a population or of an entire species. The diverse distribution of plant and animal species in the world is itself in large measure a function of climate. The only species having a world-wide distribution is ours.

Throughout the story of humanity climate has certainly been an important factor in determining large migrations, wars and changes in economic structure. Apart from this selective action or instigation to migration which we have already examined in Chapter 4, climate also has a determining action on the characteristics of a species.

Some scientists see a relationship between the gradual increase in size of the mammals during the Tertiary and the decrease in temperature during the era. According to "Bergmann's law", in fact, mammals living in cold climates are generally larger than those living in hot ones. Certainly, in our species climate has had an influence in selecting distinct genotypes more resistant to cold or to high humidity; as we have seen, the particular distribution of skin colour, hair colour; form of the nose and so on is proof of this. The factors which determine the climate of a certain region are: (a) solar radiation, (b) the astronomical position of the earth (orbit, inclination of the axis, etc.), (c) the local topography of the earth's surface and its relationship to the sea and surrounding territories. Heat emanating from within the earth has very little effect. In the course of the earth's history these factors have undergone variations, sometimes of enormous proportions. Before looking at what variations they were subjected to during the Tertiary and the Quaternary let us see how they influence present-day climate.

The upper part of the atmosphere receives an average solar irradiation estimated at 2 calories/cm² per minute. This quantity of radiation has undergone very slight variations in recent times; no data are available for the past. At present the earth's axis is inclined on the ecliptic plane at an angle of 23⁰30'. Seasonal differences and those of temperature between the equator and poles are due to this inclination. The atmosphere, with its content of water vapour and carbon dioxide, holds and diffuses the radiation energy of the sun. Seasonal thermal fluctuations are the result of the different thicknesses through which the sun's rays must pass as the inclination of the terrestrial axis varies during the year.

The unequal distribution of land and sea exerts a considerable influence on the variations (or "excursions") of the temperature of our globe. The maximum daily and seasonal variations of temperature occur at the centres of the continents (continental climate).

Temperature also varies markedly in mountain regions, diminishing with increase in altitude by 0·5 to 0·6°C for every hundred metres.

Oceans, on the contrary, reduce thermal variations (maritime or oceanic climate). Whereas, in a typical continental climate like that of Siberia, the annual thermal excursions reach values of a good 50–60°C, in the equatorial zone of the Atlantic Ocean the annual excursion is 2·5°C. Furthermore, maritime currents regulate the temperature of the oceans, and so the climate of the adjacent lands. There are also differences between the temperatures of the northern and southern hemispheres, in so far as the relationship between the seas and dry land in the two hemispheres is different. Wind is another important climatic factor. Its direction is in part due to the rotation of the earth, in part to the distribution of seas, land and high mountain relief.

Rainfall, another essential factor in the climate of a region, depends on numerous factors, the chief of which are: the winds, the degree of proximity to the sea, the ocean currents, the presence of mountain chains.

2. The Reconstruction of the Climates of the Past

To reconstruct the climate of the past we obviously cannot use thermometers and barometers; rather it is necessary to gather data of an indirect nature. Among these the most important are, first, geological and palaeontological indicators such as remains of plants and animals which have close relationships with existing forms, or at any rate have special ecological and physiological peculiarities, and, second, physical methods, such as that based on the isotopes of oxygen (^{15}O, ^{18}O).

We shall now see what the principal geological and palaeontological indicators are for the reconstruction of the climate of the Tertiary and Quaternary eras.

Hot and Cold Climates

If the degree of humidity is disregarded, there are sediments, like the laterites, which attest to hot climates. Among sedimentary rocks of marine biological origin, those which are derived from coral certainly point to climates tropical in type. However, indications that can be deduced from land plants and animals are of greater interest for our argument.

Since mammals are adaptable, finding a skeleton of a species now living in a hot climate is not in itself an indication of an equivalent temperature in the past. Of greater palaeoclimatological importance are the cold-blooded vertebrates. For them life in cold climates is very difficult, and for those of greater dimensions, as for example, the large reptiles, even the temperate zones represent inhospitable habitats. Thus reptiles are found solely near the equator and in hot humid climates. "Bergmann's law" is inversely applicable to that held valid for the mammals.

In regard to vegetation it can be said first of all that in tropical areas a more abundant number of species is found than in regions having a cold climate.

Comparing the habitats of existing species furnishes an excellent opportunity for palaeoclimatological evaluation of fossil finds. For example, by comparing the habitat of the palms and of other plants of the Tertiary with the present-day representatives of the same species it has been possible to establish the temperature of the Tertiary in Switzerland. The sequoia has a distribution area at present limited to California, whereas in the Tertiary this species was much more wide-spread (as is shown in Fig. 14.1). This is an indication that in those regions a warmer climate must have existed in the past.

Glaciers and moraines obviously constitute a very good indication of a "glacial" climate. Small isolated glaciers are found even in the tropics, but only on high mountains like Mounts Kenya and Kilimanjaro. The presence of ancient glaciers can be proved by their characteristic deposits: the moraines.

The appearance of the landscape is another important indicator for dis-

FIG. 14.1 Past and present distribution of *Sequoia* and *Metasequoia*: 1. present distribution of *Sequoia*; 2. present distribution of *Metasequoia*; 3. and 4. distribution of *Sequoia* and *Metasequoia* in the Tertiary (according to information obtained from Schloemer-Jäger, 1958).

tinguishing past glaciations, especially those of the Quaternary. The typical "U"-shaped valleys, the lakes formed by frontal moraines, the smooth, striated rocks, the erratic masses, are all indications of glaciations which existed in the more or less distant past.

Another significant marker, especially for the Quaternary, is the determination of the snow-line. The line which joins the points of the lowest altitudinal limits of the eternal snows in the various regions is called an isocheim. As it also depends on rainfall and other climatic factors, it can be seen that the limit of the permanent snow has a variable height depending on the distance of the zone from the equator. Actually, in Greenland it is at sea level, in the region of the equator at about 4000 metres above sea level.

During the Pleistocene the snow-line reached limits clearly lower than those of today. It reached levels as much as 1000 metres lower in the course of the last glaciation.

The level of the sea is greatly lowered during glacial phases, with a consequent increase in the power of erosion of rivers because of the change in level. Interglacial phases, on the contrary, provide particularly favourable conditions for sedimentation.

During the maximum extent of the glacial phases, sea level must have been at least 90 m lower than it is at present, and many of the existing ocean depths must have constituted dry land and an easy passage from islands to continents. It is estimated that, if all the glaciers now in existence were to melt, the sea level would rise by 5 metres.

In correspondence with each standstill in sea level, it is possible to identify beaches with characteristic fossil remains that furnish indications of the variations that have occurred. These data are especially important for the study of the palaeoclimate of the Quaternary and have been studied most thoroughly in the Mediterranean area. Famous studies on the Tyrrhenian coast were carried out by A. C. Blanc.

Palaeobotanical and palaeozoological indicators are of particular importance as evidence of a cold climate, especially for the Tertiary and Quaternary. Plants typical of the glacial phases of the Quaternary are: Mountain Avens (*Dryas octopetala*), Polar Willow (*Salix polaris*) and Dwarf Birch (*Betula nana*).

Today these plants are distributed only in the far north or in high mountains, while in the past they were distributed over all the central European tundra. Since the lower limit of existing tundra coincides more or less with the 10°C isotherm for the hottest months, it can be calculated that the temperature in central Europe during the Pleistocene glaciations must have been at least from 6 to 10°C lower than it is today. Another group of plants from which fairly good indications of climate can be obtained are the conifers.

The reconstruction of the vegetation of a zone is effected by means of the study of pollen-grains. Pollen grains generally preserve their cellulose walls almost unaltered; each species having its own characteristics (see Fig. 14.2).

Animals are also an excellent guide to the climatic oscillations of a region during the Quaternary. During the glacial age the tundra was inhabited by the lemming (*Myodes*), arctic fox (*Alopex lagopus*), variable hare (*Lepus variabilis*), reindeer (*Rangifer tarandus*), and musk-ox (*Ovibos moschatus*); during such periods lived large, now-extinct mammals like the mammoth (*Elephas primigenius*) and the woolly rhinoceros (*Rhinoceros thycorhinus*).

The interglacial phases were, on the contrary, characterized by the pres-

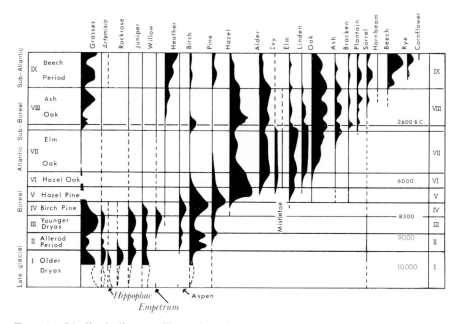

FIG. 14.2 Idealized diagram illustrating the Post-glacial pollen sequence in Jutland (from information obtained in Schwarzbach, 1946). The size of the black bands represents the percentage of pollen grains of those species found in the strata.

ence of fauna of the type existing to-day together with extinct temperate zone species, such as *Rhinoceros merkii* and *Elephas antiquus*. Some species of gastropods are also of major importance in determining climate.

Arid and Humid Climates

Geological and biological indicators also exist for determining the degree of humidity in a particular region during the past.

Accumulations of sand in the form of desert-type dunes, for example, are an indication of a dry, windy climate. In the same way the presence of the remains of xerophytic plants is characteristic of dry climates. The presence of the Irish elk (*Megaceros hibernicus*), with its four metres of branching horn, positively excludes the existence of thick forests.

The presence of dense forests and of plants with leaves rich in stomata is, on the other hand, an indication of a wet climate. With some degree of approximation it is also possible to establish for past ages other climatic characteristics such as wind-force, atmospheric pressure and the frequency and distribution of storms.

3. Seasonal Climatic Variations and those Extending Over a Number of Years

Estimation of seasonal climatic variations is useful both in itself and with the object of fixing dates. In the preceding chapter we have considered the varve method. Further information on annual climatic variations can be furnished by the woody growth-rings of plants, or by the growth lines on the exoskeletons of many gastropods and corals. The valves of the gastropods are of special interest since they contain calcium carbonate and so may be used for radiocarbon dating and for the determination of the different temperatures under which they were formed (using the ^{18}O method).

Other than annual variations, secular or cyclic variations may be revealed, extending for ten, one hundred, or many thousands of years.

For the Quaternary, as we shall see, it has been possible to recognize alternating eleven-year cycles of climate, which some scientists think are related to the eleven-year sun-spot cycles.

4. Physical Methods for the Determination of the Temperature of Past Ages

The most modern technique for obtaining information about past climate is based on an isotope of oxygen: ^{18}O. This method, devised by Urey, is based on the dependence of the relationship between ^{18}O and ^{16}O (for example in calcium carbonate) on the environmental temperature in which the oxygenated compound originated.

It is, therefore, possible to determine in this way the temperature of the sea or of the lakes in which the shells of different molluscs were formed. This information can be profitably used in the reconstruction of the climate of a given region.

5. The Climate During the Tertiary

The climatic data of the eras preceding the Tertiary can be reconstructed only superficially. At the beginning of the Tertiary, about 70 million years ago, the climate was hotter than that of today. However, during this period the temperature dropped steadily until at the beginning of the Quaternary, about one million years ago, it must have been more or less equal to the present temperature.

A higher temperature was characteristic of the entire globe at the beginning of the Tertiary. Palms grew in Alaska (lat. 62°N), and in the eastern part of Russia (lat. 55°N) during the Pliocene, and in Germany where not only the layers of the Pliocene but also those of the Miocene are rich in remains of palms.

Remains of a hot-humid flora are found in northern Japan, in New Zealand and in Chile. In the first periods of the Tertiary crocodiles infested the swamps of North America, England and Mongolia, while they do not now extend beyond northern Florida and North Africa. Termites enclosed in amber have been found in Lower Oligocene strata in Russia.

An important factor which certainly influenced the temperature of Europe during the beginning of the Tertiary was the existence of a wide passage between the Mediterranean and the Indian Ocean, the latter being by far the hotter. Figure 14.3 shows the distribution of climate and of dry land during the oldest Tertiary.

The gradual decrease in temperature during the Tertiary is demonstrated by the distribution of the flora and fauna. In central Europe there was a change from a mean annual temperature of about 21°C during the

FIG. 14.3 Map showing climate in Lower Tertiary. (Adopted from Schwarzbach, 1946.)

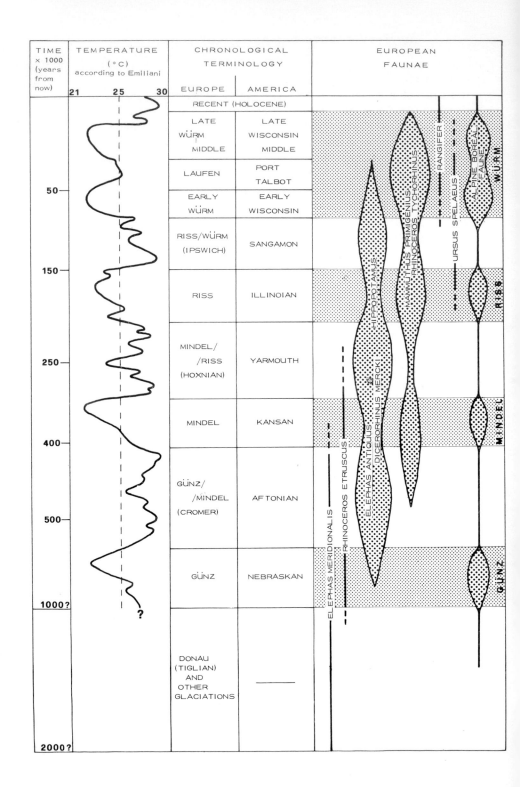

Eocene to a temperature of not over 14°C in the Pliocene. During the Pliocene there were climatic oscillations of the same type as those of the Quaternary, as is shown by sedimentary traces of glaciation and faunal and floral variations.

It is certain that during the first periods of the Tertiary the polar regions were not covered by ice-caps. The difference in temperature between the poles and the equator must have been less than it is now, and the climate all over the earth more uniform.

Gradually the temperature was lowered, the first glaciers appeared in the arctic regions and later on in high mountain ranges also.

These climatic variations were caused by major geological change. The increase in aridity in the west of North America, for example, was due to the appearance of mountain chains. Other variations were due to tectonic movements. These were sometimes on a small scale, but localized in key positions, like the closing of the connection between the Mediterranean and the Indian Oceans.

6. The Climate During the Quaternary

Although relatively brief (perhaps one million years) the Quaternary is a period that is extremely complex from the point of view of climate. Phases of glaciation were followed by temperate or hot periods in the entire northern hemisphere of Eurasia and North America. The various moraine deposits clearly demonstrate the sequence of these changes.

As was pointed out in the previous chapter, the Quaternary is sub-divided by many authors into the Pleistocene and the Holocene, the latter representing the post-glacial period. The Pleistocene can be subdivided into glacials and interglacials. The Holocene covers only the last 10,000 years (Fig. 14.4). Before going into the climatic subdivisions of the Pleistocene let us see what the climate must have been like during the glacial and inter-glacial phases.

The enormous distribution of ice during the Pleistocene in all of Europe, North America, the northern part of Asia and South America and on many mountain chains (Fig. 14.5) was caused by a diminution of the temperature.

Many geological and geomorphological traces of these glaciations exist: erratic masses, moraines, levigated rocks, glacial valleys.

The flat landscape of northern Eurasia and of North America is proof of the extension in the past of a uniform cover of ice. In mountain chains the

Fig. 14.4 Table synthesizing some of the most important chronological, climatological, and palaeontological data on the Quaternary. The absolute chronology is based on information obtained by the potassium-argon method. The curve showing temperature variation is based on data obtained by C. Emiliani on the ocean surface at low latitudes.

proof of the existence of glaciers can be found, for example, in the typical U-shaped valleys. Around the perimeters of the glaciers are found special phenomena connected with the existence of these masses of ice, such as the deposits of loess or the asymmetrical valleys.

To these geological indicators can be added biological indications such as the alternation in the distribution of the flora (which can be singled out

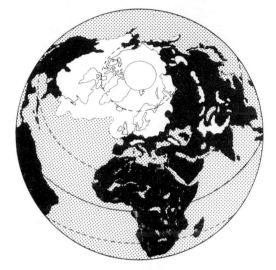

FIG. 14.5 Maximum expansion of ice-cap in the Northern Hemisphere during the Pleistocene.

effectively by means of their pollens) and of the fauna. The musk-ox, the mammoth and the woolly rhinoceros, for example, are all adapted to life in a cold climate. It can thus be established that in the northern hemisphere the temperature during the highest peak of the Würm glaciation must have been 15–16°C lower than that of today. Since the glacial zones were even more extensive during the preceding glacial phases, the temperature drop in the corresponding periods must have been even greater.

The decrease in temperature during the summer must have been somewhat different from that of the winter. In fact, the existing polar limits of plants with long stalks generally coincide with the isotherm of 10°C in July.

For example, in the eastern Alps the boundary between plants and tundra country lies at present along the July isotherm of 10°C and an annual mean of 2°C. On the basis of these and other data it can be calculated that the temperature during the maximum peak of the Würm glaciation must have been 8°C colder than that of to-day in July and 12°C colder in January. Rainfall in these glacial areas was not higher than it is now.

For the last phase of the Würm glaciation the direction of the winds prevalent in many parts of Europe can be determined by means of the study of the particular distribution of aeolian deposits comparable to dunes.

If these were the climatic conditions of the northern regions during the glacial phases, it cannot be supposed that these cold waves would have spared the regions latitudinally farther south or the now arid zones. During the glacial phases the snow-line was lowered everywhere, even on the equator, below the present-day level. According to some authors it would have descended by about 1400 m on the Alps.

At the equator, the mean annual temperature must have been more or less 4°C lower than now; that is, it must have been about 23°C. The diminution of the whole earth temperature, especially in the equatorial zones, brought, as a consequence, an alteration in the general direction of the winds, resulting in higher rainfall in the desert regions and, therefore, a remarkable reduction in them. The glacial phases of the polar regions thus correspond to so-called "pluvial phases" in the zones near the equator, concomitantly with a reduction in temperature.

There are proofs of the existence of these pluvial phases in the present-day desert regions. On the Mediterranean coast of Spain, for example, the pluvial phases—that is phases during which there was a hot-humid climate —can be distinguished by a reddish coloration; the hot-dry periods of the interglacial phases (with a climate similar to the existing one) are represented instead by calcareous deposits. In Asia at that time, the Caspian Sea was connected with the Aral Sea and the Black Sea. The Dead Sea was 400 m higher than it is at present, and the water-level was higher than it is now in the African lakes of the Rift Valley also.

During the interglacial phases, on the other hand, the climate must have been more or less the same as that of today, or perhaps slightly warmer. For this there is floral and faunal evidence as well as the geological data. Almost certainly the climate was warmer during the first interglacial phases than it is now. Proof of this is the finding of a monkey, *Macaca florentina*, in a deposit from the first interglacial in Holland. whereas the only European region in which these animals now live is Gibraltar. Many other species of animals and plants not now present in Europe but then very widely disseminated, could also be mentioned.

Since the Quaternary is of direct interest to us if we wish to follow the evolution of our species, it will be useful to examine the climatic variations which occurred in the different continents during this period. Let us begin with Europe.

During the Quaternary there were three large glacial centres in Europe: one in Scandinavia, one in the British Isles, and another in the Alps. The glacial areas of Scandinavia and the British Isles for some time extended to

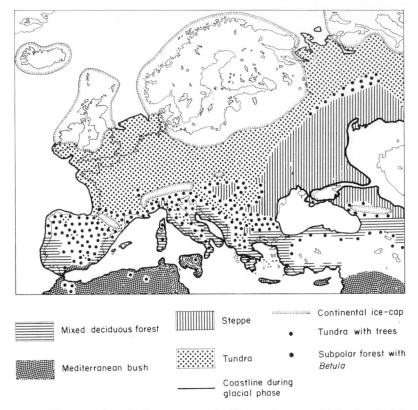

Mixed deciduous forest

Mediterranean bush

Steppe

Tundra

Coastline during
glacial phase

Continental ice-cap

Tundra with trees

Subpolar forest with
Betula

FIG. 14.6 Climate regions in Europe after the Wurm glaciation. (Following Budel and Woldstedt, from Schwarzbach, 1963.)

the point of joining to form a new, extensive glacial region (see Fig. 14.6). It is calculated that in the centre of Scandinavia the thickness of the glaciers during that epoch reached as much as 13,000 m.

In the Alps, as has been said, the line of the permanent snow was 12,000 m lower than that of today, and the entire mountain chain must have been covered by a thick mantle of ice from which the summits of the highest mountains stood out here and there. The tongues of the glaciers extended north and south at the feet of the mountain chains to form the glacial valleys.

During the Quaternary there was a succession of four principal glacial phases (followed by as many interglacial phases). They have been named after four tributaries of the Danube where they were described for the first time: Günz, Mindel, Riss, and finally the one closest to us, Würm (Fig. 14.7).

In America, too, there were different, separate centres of glaciation. The largest of these was the so-called "Laurentian", which originally arose in the plateau of north-eastern North America, to the north of present-day Quebec, in the region of Labrador in Baffin Land. Little by little it extended to the west. During the maximum glacial phase it reached the glaciers

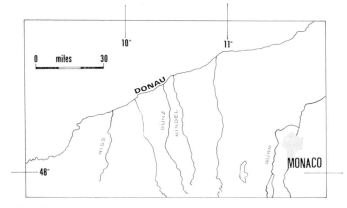

FIG. 14.7 The tributaries of the Danube, after which were named the various glacial periods of Europe.

of the Cordillera and of California forming a complete glacial blanket which covered almost all of the United States with a thickness of even more than 2500 m. At the time of the Wisconsin glaciation the ice-cap in the State of New York, for example, must have been at least 1000–1300 m deep.

In North America, as in Europe, four principal glacial phases can be distinguished: Nebraskan, Kansan, Illinoian and Wisconsin, separated by corresponding interglacial phases which are called Aftonian, Yarmouth and Sangamon (Fig. 14.4). Here as well the last glaciation determined the conformation of the present-day landscape and the formations of the Great Lakes basin.

Much less is known of the Pleistocene stratigraphy in other regions. The only thing that is certain is that in every place where careful research has been carried out, traces of different and independent glacial phases have been discovered. Correlations with the European and the North American glacial phases are still uncertain. Even less certain are the correlations between the pluvial subdivisions of southern Asia and Africa and the sequence of glaciations in Europe. The glaciations of northern Asia (Siberia) are certainly linked with the Scandinavian glacial cap.

In Africa, as in Australia, the glaciers during the Pleistocene must have been of very modest dimensions, while more extensive ones must have been present in Tasmania and New Zealand. The chain of the Andes in South

America must, in that period, have been covered as far as 26°S latitude by a uniform blanket of ice, which must also have extended over all of the plain of Patagonia.

A direct consequence of the glaciations are the eustatic movements of the sea; that is, the variations of sea-level due to the enormous quantity of

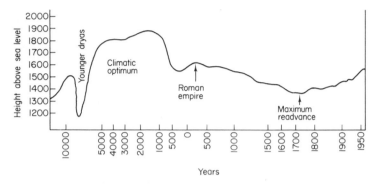

FIG. 14.8 Snow line in Norway during the last twelve thousand years. (Data modified from O. Liestol in O. Holtedahl, 1960.)

water subtracted from the sea and accumulated in the glaciers. It is estimated that during the last glaciation the sea-level was lowered about 90 m with respect to the present level. These variations in sea-level during the glacial phases, and the succession of beaches to which they gave rise, have been most thoroughly studied by Blanc in the Mediterranean area, and have been named Sicilian, Milazzian (Gunz-Mindel interglacial), Monastirian (Riss-Wurm interglacial) and, for the most recent type, Nice. A lowering of the sea-level by 80 m can completely transform the geography of a continent.

Only about 10,000 years have passed since the ice of the last glaciation (Würm) completed its retreat.

This ultimate glaciation and the following post-glacial period are particularly interesting because the final phases of human evolution took place and during this time the so-called "civilized" life of our species began.

With the retreat of the ice in Europe the climate did not improve in a constant way, but was subject to more or less wide fluctuations. It has been possible to establish the following succession of particularly well-dated events by the varve method or by radiocarbon (^{14}C) (Fig. 14.8).

As a first step there was a gradual regression of the Scandinavian glacial nucleus which led, about 10,000–9000 B.C., to a climatic optimum called the Alleröd phase. During this period the mean July temperature in central Europe must have been about 4°C colder than it is now. A new phase of climatic deterioration followed this, called the "recent Dryas", with a peak around 9000–8000 B.C. During this period temperatures in central Europe must have descended to a level 7–8°C lower than those of the present. In a more recent time the extension of the ice again began to be reduced with remarkable rapidity in all mountainous and northern areas of Scandinavia. Mean annual temperature and rainfall increased. This is defined as the post-glacial "climatic optimum" (5000–3000 years B.C.). Finally a new lowering of the temperature has led to existing climatic conditions.

Around the years 2300, 1200 and 600 B.C. there are indications of extreme dryness in all of Europe. The climate of the last 2000 or 3000 years belongs to the historical period.

Concomitant climatic changes are recorded for the North American continent.

A different type of climatic variation took place in Africa and Asia. Before the climatic optimum of which we have spoken, both in the Sahara and in the Transcaucasian regions the climate must have been more or less similar to that of a Europe free of ice. About 8000 years ago these regions began to become arid.

Events of this sort have certainly influenced the selection of human and animal types and their characteristics, as well as migrations of entire populations from one region to another.

GENERAL REFERENCES

Daly, R. A. (1934). "The Changing World of the Ice Age", Yale U. Press, New Haven.
Emiliani, C. (1963). The significance of deep-sea cores, In "Science in Archaeology", D. Brothwell and E. Higgs, (eds), pp. 99–107.
Holtedahl, O. (1960). "Geology of Norway", Norg. Geol. Unders. 208.
Sawyer, J. S. (ed.) (1966). "World Climate from 8000 to 0 B.C." Proceedings of the International Symposium held at Imperial College, London, April 18 and 19, 1966. London: The Royal Meteorological Society.
Schloemer-Jäger, A. (1958). Alttertiäre Pflanzen aus Flözen der Brögger-Halbinsel Spitzbergens, Palaeontogr. B., 104.
Schwarzbach, M. (1963). "Climates of the Past: An Introduction to Paleoclimatology", D. Van Nostrand Company, London.

NOTE ADDED IN PROOF

The recent re-evaluation of the continental drift theory will probably contribute to solving many problems concerning climate changes and species distribution and differentiation.

15

Fossil Primates

1. INTRODUCTION

In the preceding chapters we have discussed of the length of geological time and of the climatic conditions in which life on our planet developed during the final periods of the Tertiary and during the Quaternary.

We have also followed, in brief, the sequence of the various forms which from the minute *Eophippus* led to the existing stallion. By means of fossil finds it is possible to reconstruct the biological history of all animals; nevertheless, these "documents" are not always plentiful. Their abundance depends to a large part on whether or not skeletons underwent a process of fossilization and so survived to the present.

Unfortunately, we have only scanty and very fragmentary data for the primates. Since they live for the greater part in trees and in forests there is little possibility of their skeletons being fossilized and so preserved. In hot humid places rich in vegetation, the entire corpse decomposes in a short time, and the bones too decay. The teeth are the most difficult things to destroy. This is the reason, as we shall see, for the remarkable quantity of primate teeth which have survived. Teeth are, moreover, very important because they are morphologically homogeneous within the limits of each species, and stable in their dimensions once free from the alveolus. Unluckily, even teeth are not always preserved over long periods in a corrosive soil like that of tropical forests.

Most of the strata containing fossils of vertebrates are sedimentary formations laid down in shallow seas, estuaries and lagoons, to which, in many cases, the bodies or skeletal remains were transported. In such places fairly rapid sedimentation took place, and so the processes of fossilization could proceed. This happened very rarely in the case of the primates, since most of them have lived in forests. The fossil remains are reasonably abundant only for some species, for example *Macaca* and *Papio*, which live in savannah lands. The most ancient hominids also lived on the ground and in the open; later some of them found it convenient to live in caves. Only 50,000 years ago did man begin to bury his dead; from that moment onward finds of our species are fairly frequent.

2. General Conditions on Earth During the Cenozoic, and General Conditions of the Geographical Distribution of Fossil Primates

During the Palaeocene and the Eocene, the surface of the earth was much less mountainous than it is now, and tropical forests extended far into the temperate zones. This is why primate fossils of that epoch are found in areas where they no longer live. During the Oligocene the Alpine–Himalayan chain and the Rocky Mountain system began to form, and continued to rise throughout the Miocene.

The climate in the north became cooler and many forms of primates withdrew to warmer regions. During the interglacials of the Pliocene and the Pleistocene, however, the southern and western regions of Europe had a fairly mild climate, suitable for animals like the primates that lack defences against extremes of climate.

Certain phases or events in the story of the primates can be explained by the formation or the elimination of "bridges" between continents and islands. Madagascar, for example, remained separated from the African continent during all of the Jurassic, about 160 million years ago, when the mammals began to evolve. One hundred million years later, during the Eocene, the ancestors of the existing lemurs reached this African island, probably by a temporary bridge of dry land. From that epoch onwards no animal that could compete in the arboreal life of the lemurs arrived in Madagascar until the island was reached by man about 2000 years ago.

Africa and Eurasia were connected intermittently across the zone of the Suez. At different times these temporary connections permitted the passage of some forms of catarrhine monkeys from Eurasia to Africa, and *vice versa*. From the Cretaceous to the middle of the Eocene there was a passageway over the Bering Strait, across which the American continent could be reached easily; it was then covered with water until the Pleistocene when man and other animals could cross it once more. After that water again became an obstacle.

During the Cretaceous and early Palaeocene the Isthmus of Panama joined North and South America, permitting the prosimians to move south into the neotropical forest where, as we shall see, they gave origin to the diffusion of the platyrrhine monkeys and proliferated in a variety of forms. In the meantime, the prosimians of North America became extinct. The isthmus remained submerged until the late Pliocene when it re-emerged and acquired the conformation of today.

Because of this re-emergence many new species of animals invaded South America, which led to the extinction of numerous species unable to compete with these new arrivals. The competition did not, however, interfere

FIG. 15.1 Distribution of fossil primates in the world.

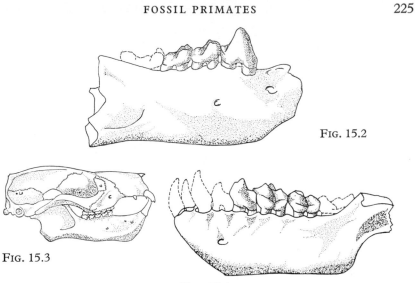

FIG. 15.2

FIG. 15.3

FIG. 15.4

FIG. 15.2 *Phenacolemur frugivorus* (Palaeocene, North America). Ramus of the mandible.
FIG. 15.3 Skull of *Plesiadapis*: left lateral view.
FIG. 15.4 *Anaptomorphus aemulus*: left lateral view of the mandible.

with the various animals that were largely arboreal. On the contrary, some of them, moving from south to north, re-invaded part of the territories to the north of the isthmus, where they now live.

During the Cenozoic era Australia had no dry-land connection either with Africa or Asia, so that no primates reached it until man arrived during the late Pleistocene.

3. THE FIRST FORMS OF THE PROSIMIANS AND THEIR MULTIPLICATION

During the first period of the Tertiary era (from the Palaeocene to the middle Oligocene) among the forms of mammals in the course of development, four were in the process of specializing their dentition for a particular way of life.

They were the multituberculates, the Primates, the Rodents and the Lagomorphs. The first to develop were the multituberculates, which reached their maximum development during the Palaeocene but became extinct in the Eocene. The primates (represented by the first forms of prosimians), although having to compete with the multituberculates, rodents and lagomorphs, soon evolved into many diverse forms.

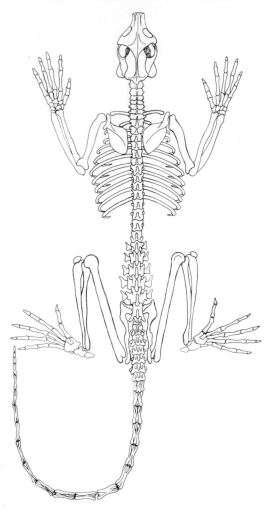

FIG. 15.5 Reconstructed skeleton of *Northarctus tyrannus* (Middle Eocene, North America).

Fossil remains which can be ascribed to about 55 genera of prosimians have been discovered in the deposits of the early Tertiary. (Fig. 15.1). They can be divided into six families: Tupaiidae, Phenacolemuridae, Plesiada-pidae, Notharctidae, Adapidae, Anaptomorphidae (Figs 15.2, 15.3, 15.4, 15.5, 15.6).

Before the middle Oligocene three of these families (Phenacolemuridae, Plesiadapidae and Notharctidae) became extinct, perhaps because they

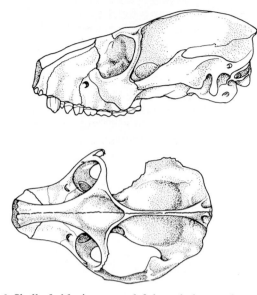

FIG. 15.6 Skull of *Adapis magnus*: left lateral view; and, superior view.

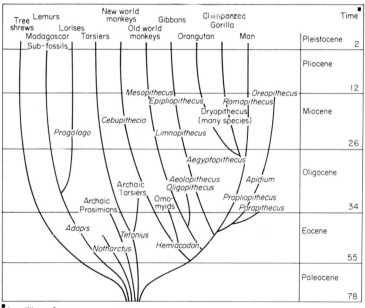

In millions of years

FIG. 15.7 Phylogenetic tree for the Primates based on fossil remains.

were over-specialized and were in competition with the rodents and lago-morphs. The Adapidae and the Anaptomorphidae, having no inclination toward rodent characteristics, owed their continued existence to their lack of specialization. *Daubentonia*, which is the only living prosimian with the specialization of a rodent, has probably acquired this differentiation only recently. Remains of Adapidae have been found both in Europe and in America. They were primates similar to the lemurs (and for this reason, sometimes called "protolemurs"), but were not specialized. They represent the ancestral forms of the living Lemuridae and Lorisidae, or their close relatives, but it does not seem that they are in any way connected with the forms that led to the higher monkeys and to man.

The Anaptomorphidae, with forms in both the Old and New Worlds, were tarsiers of non-specialized types which probably had already been separated from the group of "protolemurs" (Adapidae). It is thought that all the existing forms of monkeys, anthropoid apes and man originated in this family before the ancestors of the tarsiers became specialized. A diagram of this evolutionary tree is given in Fig. 15.7.

4. Fossil Forms of the Platyrrhines

In Chapter 11 the origin of the Platyrrhine monkeys was mentioned in another context. The classic theory suggests that the South American Monkeys originated from a group of extinct North American Prosimians (Omomydae) whose representatives migrated from North to South America and there differentiated into many diverse forms. The recent re-evaluation of the continental drift theory however makes it possible to speculate about direct migration, of some species at least, from Africa directly to South America up to the end of Eocene. The hypothesis of a direct migration was originally postulated by Lavocat in 1969.

The most ancient fossil form of Platyrrhine was found in the lower Miocene strata of Patagonia. Other and more recent fossils are described with the names, *Neosaimiri* and *Cebupithecia*.

5. Fossil Forms of the Catarrhines

Contemporaneously with the evolution of the platyrrhine monkeys, the Old World monkeys or catarrhines, the anthropoids and the hominids evolved in different forms from one or more species of Anaptomorphidae. Fossil remains that definitely connect the Old World monkeys with those of the New World or with the various forms of existing prosimians are not known. On the contrary, the New World monkeys, for the reasons given above, seem to have followed a completely independent course of evolution. The fossil remains of tarsiers known up to the present seem too peculiarly specialized to be direct ancestors of the Old World monkeys. The remains of *Necrolemur* (Fig. 15.8) and *Microchoerus* are the only forms discovered to date which present evolutionary characteristics leading in the direction of the catarrhine monkeys.

It is difficult to say whether the Cercopithecidae represent a stage in the evolution of the Hominidae or whether the two groups have evolved independently. Other than the differences of posture and of locomotion there are, as we have seen, many other major differences between them. The Cercopithecidae, for example, have a placenta with two discs, while that of the manlike apes and man has only one. Up to a few years ago a common origin for, or at least a connecting link between, the Cercopithecidae and Hominidae seemed to be attested by two small and very ancient fossil mandibles. One, which has been given the name *Amphypithecus*, was found in Burma in a deposit attributed to the Upper Eocene; the other, *Parapithecus* (Fig. 15.9) was found in the Fayum in Egypt, in a soil attributed to the lower Oligocene. However, these finds have come under some criticism.

The first fossil that can be attributed with certainty to an ancestor of the Cercopithecidae was found in eastern Africa in an upper Miocene soil, and has been named *Mesopithecus*. (Fig. 15.10). Other animals apparently of the same genus or a similar one have been found in Germany, in Czechoslovakia and in Iran, in strata of the late Miocene or early Pliocene. Of these, the most complete is that found at Pikerni, in Greece and studied by Gaudry. It has been attributed to the subfamily Colobinae, since the cranium, the mandible and the teeth resemble those of the Asiatic representatives of this group. This masticatory apparatus is clearly of the vegetarian type. However, the rest of the skeleton is less specialized than that of the living Colobinae and more similar to those of the macaques and baboons.

The femur is longer than the humerus. The ischium has a wide, hard area which suggests the existence of the ischial callosities typical of the macaques and baboons. For these and other reasons it can be assumed that this monkey lived chiefly on the ground. This find can, therefore, be considered as being intermediate between the two groups of living Cercopithecidae (Cercopithecinae and Colobinae), although approaching more closely to the Colobinae.

Mesopithecus survived in a number of forms until the Pliocene in India, and the early Pleistocene in France.

Up to the present, remains of fossil Cercopithecinae which can be attributed to periods earlier than the Pliocene have not been found. In the Pliocene diverse species of macaques (*Macaca prisca* and others) appeared in France, Holland and Italy. Other fossil remains of macaques have been

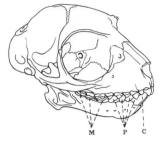

FIG. 15.8 Skull of *Necrolemur antiquus*: right lateral view. (M=molar P=premolar, C=canine teeth).

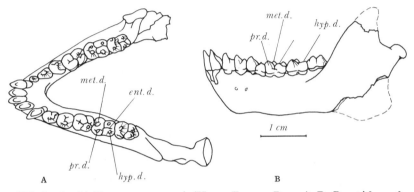

A

B

FIG. 15.9 A. *Amphipithecus mogaungensis* (Upper Eocene, Burma), B. *Parapithecus fraasi* (Oligocene), mandible with teeth from superior view. (metd. = metaconide; hyp d. = hypoconulide; pr.d = protoconulide; ent.d. = entoconide).

found in other places in Europe, Indochina and China, in Pliocene soils. Recently numerous fossil remains of a form certainly an ancestor of *Papio* and *Theropithecus* have been discovered in levels of the very early Pleistocene in different localities of South and East Africa. To these fossil remains the name of *Simopithecus* (Fig. 15.13.) has been given.

6. THE PHYLUM OF THE GIBBONS

The most ancient exemplar of the hominids found to date has many points of similarity with present-day gibbons, and probably specialized groups of the Hominoidea in fact originated from them. There are at least three skeletal fragments which throw light on the evolution of the gibbons.

The first is *Propliopithecus maeckeli* represented by a small, but almost complete, mandible. It is about two-thirds the size of that of a living gibbon, and, from this, it is deduced that this primitive hominid must have been only slightly bigger than a large cat. The mandible is V-shaped and,

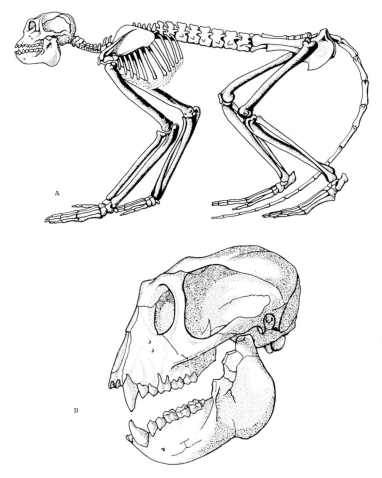

FIG. 15.10 *Mesopithecus pentelici*: A, skeleton; B, left lateral view of the skull.

therefore, very similar to that of a prosimian. The ascending rami are higher than those of the existing gibbon, which suggests a longer face. The molars have five cusps and the premolars two. Although the canines have been damaged, presumably they were of the same length as those of the present-day gibbon; the first premolar is not completely divided; the incisors are missing. In general, the dimensions of the mandible are massive. *Propliopithecus* is certainly an ancestor of the existing gibbons, but it is possible that it was also concerned in the evolution of the anthropoid apes and so of man.

Another find with "Gibbonoid" characters was found on the island of Rusinga in Lake Victoria (Kenya) in strata that can, as a complex, be dated to the lower Miocene. It has been given the name of *Limnopithecus* (Fig. 15.11). The find consists of fragments of the mandible, teeth and long bones. The canines are shorter than those of existing gibbons and the lower pre-molars less specialized; the incisors are slightly smaller and the mandible more robust. The skeleton of the limbs has many characters intermediate between those of the Cercopithecidae and those of the Hylobatidae. The lower limbs are shorter than those of the living gibbon and the whole shoulder-girdle is less specialized for brachiation.

In the middle and upper Miocene another form ancestral to the gibbon lived in Europe (Austria). It has been assigned the name *Pliopithecus* (Fig. 15.12). Its interorbital space is larger and the nasal region wider, and also the body skeleton displays many resemblances to that of the existing gibbons, whether in details of the pelvis or characteristics of the vertebrae. The sternum is wide and flat like that of the brachiators and of man, but the clavicle is S-shaped like that of the anthropoid apes. The trunk and limbs have characters in common with the gibbons, the Cercopithecines and the other anthropoid apes. From these data it is clear that *Pliopithecus* was not entirely a brachiator. The remains of *Limnopithecus* and *Pliopithecus* have thus demonstrated the presence of forms ancestral to the gibbons in

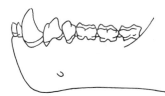
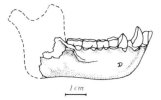

FIG. 15.11 *Limnopithecus macinnesi*: a left lateral view of the mandible showing external features.

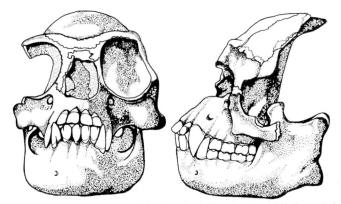

FIG. 15.12 Skull of *Pliopithecus* found in Neudorf (Czechoslavakia): frontal view and left lateral view.

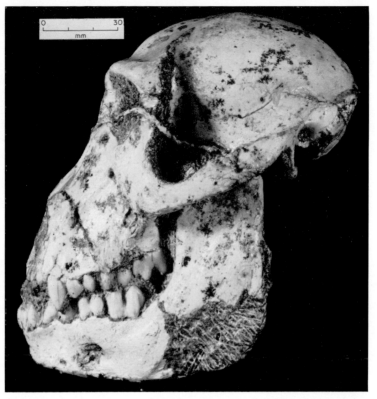

FIG. 15.13 Left lateral view of *Simopithecus darti*. Note the tall ascending ramus of the mandible and the comparatively short occipital region of the braincase. (Courtesy of Maier).

Europe and Africa during the Miocene. As far back as the Miocene, therefore, the gibbons split off from the group of other catarrhines and began to be differentiated from the ancestors of the true apes and man.

7. THE ANCESTORS OF THE ANTHROPOID APES

Moving backwards in time, unfortunately we can still say nothing about the Pleistocene ancestors of either the gorilla or the chimpanzee. Only for the orangutan do we have a few teeth that illustrate his immediate pre-history. On the other hand, the evolutionary line which led to the existing anthropoid apes is very well marked out during the last periods of the Tertiary; the Miocene and Pliocene (Fig. 15.14).

During the Miocene, the period in which the first ancestors of the gibbons appeared, the ancestors of the chimpanzee (*Pan*) and the gorilla (*Gorilla*) also appeared, probably derived from the same original stock. Remains of

FIG. 15.14 Map showing the distribution of the Dryopithecinae (Miocene and Pliocene). The fossils are represented on the map by black dots.

possible Tertiary ancestors of the orang (*Pongo pygmaeus*), on the contrary, have not been found to date. Some authors classify these forms ancestral to the apes as a subfamily in themselves—the Dryopithecinae—and call their existing descendants Ponginae; but rather than being distinct forms they are linked through descent, and so the terms Dryopithecinae and Ponginae must, in effect, indicate successive stages of evolution.

8. *Proconsul*

The most ancient of these forms is *Proconsul*, which has an African distribution. It is now considered to be a subgenus *Dryopithecus*, following the proposal of E. Simons. Remains of these animals were found on the island of Rusinga in Lake Victoria, in the same deposit in which the fossils of *Limnopithecus* were found. These remains can be differentiated by size into three different species: *Dryopithecus (Proconsul) africanus*, a little larger

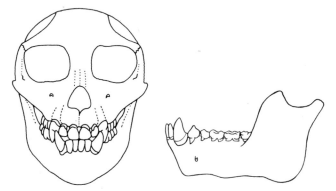

FIG. 15.15 Frontal view of the skull of *Proconsul africanus*. At the right a diagram of the mandible of this fossil (left lateral view).

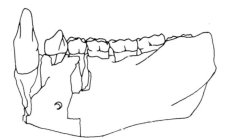

FIG. 15.16 Left lateral view of the external features of the mandible of *Dryopithecus fontani*.

than a gibbon; *Dryopith.* (*Proconsul*) *nyanzae,* the size of a chimpanzee; and *Dryopith.* (*Pronconsul*) *major,* the size of a gorilla. Except for their dimensions they are very similar morphologically.

Dryopithecus (*Proconsul*) *africanus* is the best known, because of the abundance with which its remains have been found (Fig. 15.15). On the whole it appears to have been an arboreal, quadrupedal brachiator. The head is roundish. The forehead forms an angle of 55° with the plane passing through the orbital cavity and the auditory meatus (in the gorilla this is a 35° angle and in man about 90°). The cranial capacity has not been measured directly but can be considered as being intermediate between that of the living gibbon (about 200cc) and that of the chimpanzee (about 400cc). The frontal part of the endocranial cavity displays many characteristics more similar to those of the catarrhines than to those of the great apes. The inter-orbital distance is large and the orbits are disposed slightly to the sides, from which it is presumed that these animals did not have completely stereoscopic vision. The mandible is rather small and is still slightly V-shaped. The muscle attachments are not massive as in the lower monkeys.

The incisor teeth have the dimensions of human ones, the cranium is larger, but not like that of the living apes, and on the maxillary part of the face there is a small depression (*fossa canina*) which is found in man, but not in the apes. The first lower premolar has a divided cusp, like that of the existing anthropoid apes. The upper limb is of special interest in solving the problem of possible brachiating activity of these animals. The humerus, the radius and ulna, the carpal and metacarpal bones and the phalanges, although they have some of the characteristics of the arboreal quadrumanous primates, clearly display the characteristics of good brachiators. On the other hand, none of the typical characteristics of the terrestrial quadrumanous monkeys, like the *Macaca* and *Papio*, have been found.

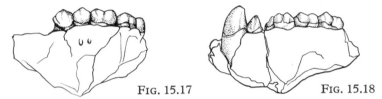

FIG. 15.17   FIG. 15.18

FIG. 15.17 *Sivapithecus indicus*: A, right lateral view of the maxilla.
FIG. 15.18 Maxilla of *Ramapithecus brevirostris*: right lateral view.

9. DRYOPITHECINAE OF EUROPE AND ASIA

At the same time as the African forms discussed above, many similar types lived and became differentiated in Europe and Asia. Further, the name of the entire group is taken from the first of these finds, a mandible discovered in France in 1856 in middle Miocene strata, *Dryopithecus fontani* (Fig. 15.16). Remains of *Dryopithecus* have been found as well in various other parts of Europe, where this genus survived until the end of the Pliocene, in southern Asia, and in lignate strata in China. In the first decades of this century a large number of teeth and mandibles of Dryopithecinae were found in fossiliferous deposits in northwestern India (the Siwalik hills). These remains, formerly classified into different genera: *Sivapithecus*, *Sugrivapithecus*, *Bramapithecus*, *Ramapithecus*, and *Paleosimia*, can all be attributed to the lower Pliocene. Of these, at present the group led by *Sivapithecus* (Fig. 15.17) is ascribed as a subgenus to the genus *Dryopithecus*, (as, as has already been said, happened to the African *Proconsul*), while *Ramapithecus* (Fig. 15.18) and the others little akin to *Sivapithecus* are grouped in a genus by themselves (*Ramapithecus*).

The fossil which displays most "hominoid" characteristics is that classified as *Ramapithecus*. It, or something very similar, is certainly at the base of the evolution of the Hominidae, whose centre of origin has been placed by some workers in eastern Asia.

10. *Gigantopithecus blacki*

The remains of *Gigantopithecus* can be classified with the group of Asian *Dryopithecinae*, even if it differs in some characteristics and is also more recent (Pleistocene) (Fig. 15.19).

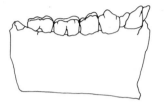

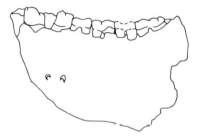

FIG. 15.19 Mandible of *Gigantopithecus*: young male (facial and occlusal surfaces); adult male (occlusal and facial surfaces).

It is certainly one of the largest primate forms that has existed up to now. Originally it was described on the basis of a few teeth found by Von Koenigswald in a Chinese pharmacy. Only recently (1956) has a mandible been discovered in strata of the ancient and middle Pleistocene of southern China. The canines of *Gigantopithecus* are smaller than those of any other ape and, apart from its enormous size, it appears to be more similar to man than to the living apes. Still more recently two other incomplete mandibles have been collected which can be ascribed to the same type and which the discoverers, the Chinese anthropologists Pei and Li, attribute to the Villafranchian; that is, to a period preceding the middle Pleistocene.

11. *Oreopithecus bambolii*

Another fossil find which could assume great importance for its characteristics in part hominoid and in part cercopithecoid, is *Oreopithecus bambolii* (Fig. 15.20).

FIG. 15.20 The skeleton of *Oreopithecus bambolii* found near Baccinello (Grosseto) in lignite deposit the 2nd of August, 1958. The bones have been cleaned and restored by Annalisa Berzi of the Palaeontological Museum in Florence, Italy, where the fossil is now kept.

The first remains of *Oreopithecus* (an incomplete juvenile mandible and fragments of other bones) were discovered by Cocchi, about 1871, in the lignites of the Montebamboli mine, from which the fossil takes its name. Other remains were found at Ribolla and recently (1954, 1958) at Baccinello, all three sites being near Grosseto. Except for the last skeleton which was complete or almost so, the finds generally consisted of a number of mandibles, three jaws, the proximal end of an ulna and of a femur.

Gervais, who studied the first finds (1872), described them as those of an anthropoid ape; later other workers revealed, especially on the third molar, some similarities to the Cercopithecinae and classified the *Oreopithecus* among them. Recently Hürzeler has replaced it in the evolutionary line of the hominids, thus moving it up to the lower Pliocene.

It is possible that this form has no connection with Dryopithecinae, nor with the other forms which we have described, but that it represents a distinct evolutionary branch which split off from the main group before the Miocene.

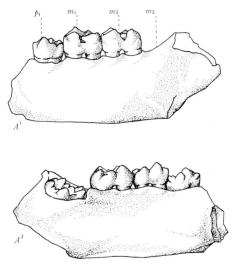

FIG. 15.21 *Apidium phiomens* (Oligocene) Fragment of mandible with P1 to M3, from the external lateral view (A1), and internal lateral view (A2).

Apparently forms similar to *Oreopithecus* have been found in Bessarabia also. The teeth of these remains, furthermore, display many resemblances to the remains of those of *Apidium* found in the Oligocene strata in Egypt (Fig. 15.21).

Since in the Oligocene deposits of Egypt, other than the remains of *Apidium*, there have also been found those of *Parapithecus* (the much-discussed form which has already been mentioned), of *Propliopithecus* (claimed precursor of the Pongidae) and of *Moeripithecus* (the possible precursor of the Cercopithecines), the three groups of Cercopithecoidea, of Hominoidea and of Oreopithecoidea must already have been distinct as independent groups at that time. While Cercopithecoidea and Hominoidea still have living representatives, the Oreopithecoidea disappeared in the lower Pliocene.

GENERAL REFERENCES

Clark, W. E. Le Gros (1965). "History of the Primates", British Museum (Natural History), London.

Gavan, J. A. (ed.) (1955). "The Non-Human Primates and Human Evolution", Wayne University Press, Detroit.

Genet-Varcin, E. (1963). "Les Singes Actuels et Fossiles", N. Boubée et Cie, Paris.

Lavocat, R. (1969). La systématique des Rongers hystricomorphes et la dérive des continents. *C.R. Acad. Sci. (Paris)* **269**, 1496–1497.

Maier, W. (1970). Neue Ergebnisse der Systematik und der stammesgeschichte der Cercopithecoidea. *Z. Saugetierkunde*, **35**, 193–214.

Maier, W. (1972). The first complete skull of *Simopithecus Darti*, from Makapansgat, South Africa and its systematic position. *J. Hum. Evol.* **1**, 395–405.

Schultz, A. H. (1969). "The Life of Primates", Weidenfeld and Nicolson, Hampshire.

Simons, E. L. (1963). A Critical Reappraisal of Tertiary Primates *in* "Evolutionary and Genetic Biology of Primates", J. Buettner-Janusch (ed.), Vol. 1, pp. 65–129.

Simons, E. L. (1964). The Early Relatives of Man, *Scient. Am.* 1964.

Simpson, G. G. (1965). "The Geography of Evolution", Churchman, Philadelphia.

16

Hominoid and Hominid Fossils

1. HOMINOIDS AND HOMINIDS

Two fundamental characteristics distinguish the hominids from their nearest relatives, the Pongidae: their posture and their teeth. In fact, hominids, by definition, have an erect carriage and walk with their arms free of the ground; their canines are small and do not protrude beyond the line of occlusion of the other teeth, nor do they have a diastema between the canine and the first upper premolar. The questions that must be answered are when and how this group of ancestral forms was differentiated and each began to acquire its own particular characteristics. Opinions regarding these problems are still very much at variance. It has recently been suggested that *Ramapithecus* should be recognized as the possible Tertiary ancestor of the Hominids. Many characteristics of this fossil, especially dental ones, would make this seem probable. The teeth are almost flat, with few ridges in the enamel; the canines small; the anterior premolar larger than the posterior and bicuspidate, without pithecoid ridges. Its dental arcade is, moreover, paraboloid, which is a characteristic of the Hominoids.

According to recent opinion the Hominoidea must have sub-divided into three groups during the Tertiary (at the beginning of the Miocene). One of these, which was different from all the modern anthropoid apes and is now extinct, was derived from a completely unspecialized form. The branch that has given origin to the existing anthropoid apes (Pongidae) may itself have originated from this group, with a specialization of the canines and a better adaptation to arboreal life. The Hominidae may have originated from the same group, following, however, a different line of specialization; that is, the reduction of the canines and the acquisition of an upright posture for adaptation to terrestrial life. This separation must have taken place in the upper Miocene or lower Pliocene, about 10 or 15 million years ago.

2. THE AUSTRALOPITHECINES

This name designates a group of fossil remains that are certainly in the line of Hominid evolution. It is a rather numerous and heterogeneous group of fossils found from 1924 onwards in Austral Africa (hence the name) or more precisely in the Transvaal, almost all close to Johannesburg (Fig. 16.1). The trend is to associate at least the most ancient of the Hominid remains found in Tanzania (Olduvai) and called *Zinjanthropus*, as well as recent findings in the Omo Valley in Ethiopia with these.

The main places of origin of the material found to date, and its composition, are as follows:

Australopithecus africanus studied by Dart, 1925; discovered at Taung. A juvenile cranium.
Plesianthropus transvaalensis—Broom 1936—Sterkfontein. Fragments of cranium, 141 teeth, different bones of the skeleton, including the pelvis.
Australopithecus prometheus—Dart 1947—Makapansgat. Fragments of cranium, 28 teeth, various other bones including the pelvis.
Paranthropus crassidens—Broom 1949—Swartkrans. Fragments of a number of crania, two of them with an obvious sagittal crest, about 270 permanent teeth and a complete mandible.
Paranthropus robustus—Broom 1939—Kromdraai. Various fragments, 6 milk teeth. The material is ascribed to about 70 individuals.
Zinjanthropus boisei—Leakey 1959—Olduvai, Tanzania (East Africa). Cranium almost adult in form, with sagittal crest. Cranial capacity of about 600cc. Face gross with pithecoid-type nostrils, very large premolars, canines and incisors disproportionately small.

The geological age of the finds cannot be defined with certainty due to the difficulty of establishing parallels with European chronology. It would seem, however, that they should be placed in the first phase of the Pleistocene, which at present is considered to have begun a couple of million years ago. The most ancient *Zinjanthropus* has been dated by the potassium-argon method to 1,750,000 years ago.

If the material found to date is taken as a whole the following general characteristics can be deduced:

The cranium is small, both in the absolute and in the relative sense; the (estimated) capacity varies from 500 to 650cc, with a possible maximum of 750cc for one of the crania of *Paranthropus crassidens*.

FIG. 16.1 The 11 African Pleistocene sites which have yielded fossils positively or probably identified as Australopithecine remains. Five sites are in the Republic of South Africa, three in the Republic of Tanzania, two in the republic of Kenya and one in Ethiopia.

The face is large with marked prognathism; the teeth are sometimes larger than those of a gorilla (*P. crassidens*); very rarely does the length of the canine exceed the plane of mastication.

Morphologically, the hipbone begins to resemble that of humans. The wing of the *ilium* is wide as in man, while in the anthropoids it is much narrower; moreover, in the anthropoids, it lies on an almost perpendicular plane with respect to the opening of the acetabulum, in the Australopithecines it lies on a plane almost parallel to it—a condition closely approaching that of man. Another important characteristic of the pelvis of the Australopithecines is the presence, as in man, of an anterior-inferior iliac spine (which has connections with the ilio-femoral ligament and thus with erect posture), whereas this structure is absent in the anthropoid apes.

Since the morphology of the hip bone is closely related to posture, it is probable that the Australopithecines were endowed with upright carriage. The taxonomic position of this group of finds has been the subject of many disputes. While some place them among the Pongidae, others consider them to be much closer to man. Without doubt they are closer to man than to any living anthropoid ape. Heberer, in order to underline this fact, has appositely coined the name "Prehominine", but this term contains within itself the idea that these might be the true ancestors of man, which, as we shall see, is not true.

The Australopithecine have many pithecoid characteristics in the cranium if not in the teeth. On the other hand, many characteristics, such as an erect or semi-erect posture, place them closer to man. They appear to be an intermediate type constituting a distinct and fairly compact group of hominids, which must be considered as an extinct side branch of the evolutionary tree which has produced man.

The problem is further complicated for many workers by the fact that traces of industry, and possibly of carbon, have been found along with some Australopithecine remains. In that case, notwithstanding their largely pithecoid characteristics should the Australopithecines be considered as "men" or as "anthropoid apes"?

Admitting that the remains of carbon are truly from wood burned by these beings, and that the pebbles found have really been flaked by them (which, even if not proved, seems highly likely), the problem is simply one of altering an assumption based on an anthropocentric psychology. Up to the present it has been maintained as a statement of fact that monkeys could use, in a limited fashion, pre-existing implements, but that the manufacture of these, for either an immediate or later use, was a faculty of man alone; that is, it was thought that only man had the capacity to move through time mentally, but that is a purely anthropocentric axiom, of a limiting type.

The capacity to exercise a measure of dominion over nature, to manufacture artifacts, to move freely in space and in time, are not in fact traits restricted to *Homo sapiens* alone, but are things of which many animals, even those outside the order of Primates, are capable, even if to a lesser

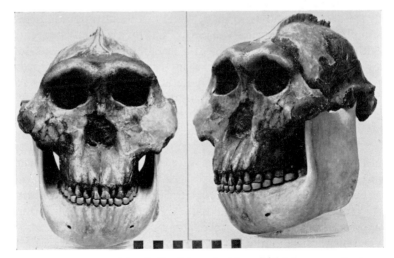

FIG. 16.2 Reconstruction of the skulls of *Australopithecus boisei* the most robust among the Australopithecine Bed I, Olduvai Gorge (courtesy of P. Tobias, 1968).

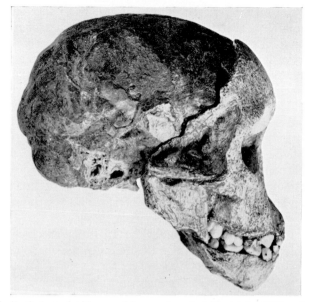

FIG. 16.3 Skull of *Australopithecus africanus*, found at Taung in 1924 and first reported by Dart (courtesy of P. Tobias, 1968).

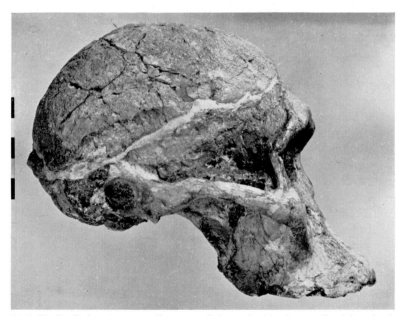

FIG. 16.4 Skull of the more gracile type of Australopithecine at Sterkfontein (st 5), discovered by R. Broom and J. T. Robinson on the 18th of April, 1947 (courtesy of P. Tobias, 1968).

degree. Consciousness of oneself, introspection, and, hence, auto-criticism, that is, the possibility of self-perfection, are the true characteristics of Man. Recent research on psychology of the primates, particularly of the anthropoid apes, is revealing remarkable intellectual abilities in these animals as well.

As has been pointed out, the Australopithecines, even if they represent a collateral branch, must certainly be considered as a stage on the way to the human condition in which one of the essential characteristics of man the ability to stand erect, had either already been developed or was being so. It is this which strictly governs the evolution of intellectual capacity; Aristotle had already divined this when he affirmed that erect posture was the "first condition" of thought.

The Australopithecines certainly demonstrate that of the two particular characteristics specific to man, upright carriage and large cranial capacity, upright carriage was realized first phylogenetically.

To clarify the taxonomic position of the Australopithecines with respect to the Hominids it might be useful at this point to consider the evolution of another group of mammals: the elephants.

There are two distinct species of elephants: the Indian and the African, with a totally different geographical distribution. Linnaeus originally included both in the genus *Elephas*, for they have many characters in common. They have almost the same dimensions, both are endowed with tusks, and they have the same bodily conformation. However the Indian elephant has smaller ears than the African.

Detailed study of the skeleton and in particular of the teeth has demonstrated that, on the contrary, the two types represent different stages of evolution, and that they must be attributed to two different genera, *Loxodonta africana* and *Elephas indicus*. The genus *Loxodonta* is characterized by having shorter teeth, with a smaller number of prominences (a maximum of about 12 as compared with the 24 of the last upper molar and the 27 of the last lower molar of the Indian elephant). The Indian elephant is, therefore, much more specialized. Notwithstanding the different degree of specialization they share a common ancestor in the *Archidiskodon* of the lower Pleistocene.

These differences have thus been realized in a period of 500,000 years. An equivalent period of time may also have elapsed between the evolutionary differentiation of the forms of *Australopithecus* and those of *Pithecanthropus* from a common Pliocene ancestor.

3. *Pithecanthropus*

The Dutch doctor Dubois gave the name "*Pithecanthropus*" to a primate of which he had found the first fossil remains in 1891 near Trinil, on the island of Java (Fig. 16.5). In this primate, as in the Australopithecines, pithecoid and human characters appeared in association. Excavations, for many years unfruitful, in 1936 under the direction of Von Koenigswald led to the discovery of other morphologically rather heterogeneous finds, which can be distributed provisionally in three species: *Meganthropus paleojavanicus*, *Pithecanthropus modjokertensis* and *Pithecanthropus erectus*.

The first comes from the deposits of Trinil, the other two from the deposits of Djetis, both deriving from the lower Pleistocene. Of the three, especially if their morphological characters are taken into consideration, *Meganthropus* is probably the most ancient, followed by *Pithecanthropus robustus* and *Pithecanthropus erectus*, going back to the Mindel glaciation.

Meganthropus (Fig. 16.6) is represented by two fragments of mandibles of which the second is as high as the mandible of a gorilla. The three teeth present are enormous. They display a human structure, although the sequence of size of the premolars and molars is pithecoid in type. The canine is missing but the size of the alveolus suggests a very large tooth. These fragments of mandible display a remarkable resemblance to analogous remains of some Australopithecines (*Paranthropus*).

Pithecanthropus robustus is represented by the posterior part of a maxilla, by other small fragments and by numerous teeth. The cranium on the whole has some very primitive characters, even if the size appears to be greater than that of *Pithecanthropus erectus*. The occipital torus is enormous and continues in a remarkable sagittal crest. The anterior facial profile is concave as in the orangutan, with marked prognathism. The canines extend slightly beyond the other teeth and are preceded by a diastema. The other teeth are very primitive, very large, taurodonts. For *Pithecanthropus erectus* (Fig. 16.7) which is represented by more abundant remains from both sexes, unfortunately only skullcaps have been found, and no complete crania. The skullcaps have rather small dimensions, especially the width and height, as compared with human crania. The endocranial capacity is estimated to be about 900cc in the male and 750cc in the female (*Pithecanthropus II*). It is therefore intermediate between the average for the Australopithecines (600cc) and the lowest average of existing populations (1200cc).

The cranium considered from above shows peculiar characters: a marked retro-orbital narrowing and a sagittal ridge which does not have the value it has in the gorilla, that is, for the insertion of the temporal muscles, but is solely morphological in character (a crest). This peculiarity is also found in *Sinanthropus* (see below), and in some primitive groups such as the Fuegians and Eskimos. From the front and side the cranium displays a strong, continuous supra-orbital bar, the supra-orbital torus, below a low, receding forehead like that of the chimpanzee. The posterior aspect reveals the existence of an "occipital torus" which seems to continue, as happens in

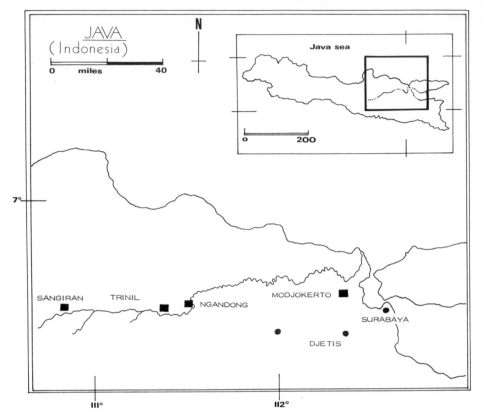

FIG. 16.5 Map showing location of Hominid finds in Java.

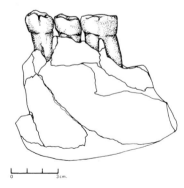

FIG. 16.6 Fragment of mandible of *Meganthropus II* (by courtesy of G. H. R. von Koenigswald, 1941).

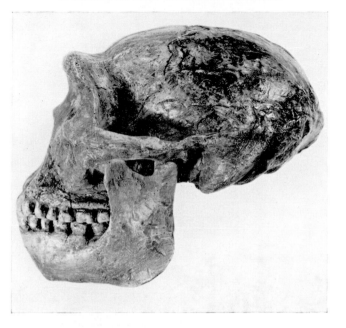

FIG. 16.7 *Pithecanthropus mo djokertensis*, new reconstruction. Capacity 750cc, male skull Sangiran, Java (by courtesy of G. H. R. von Koenigswald).

the anthropoid apes but not in present-day man, in the supra-mastoid crest. Sergi attributes great importance to this character.

The mandible is represented by two fragments. It is very large, lacking a chin, and predominantly pithecoid in character. The teeth are large, taurodonts, and their dimensions increase progressively from the first to the third molar. The skeleton of the limbs is represented by the left femur and five fragments, all perfectly human in type and comprised within the limits of dimensional variability of existing human beings. The entire femur suggests a stature of 165–170cm, which would seem to be too tall.

It would appear that these three forms, although displaying marked differences, can be placed in the same evolutionary line. *Meganthropus*, also called Pre-*pithecanthropus*, would be the most ancient and most primitive form; the others gradually progressing. No implements of any industry are known to be securely associated with remains of *Pithecanthropus*. In the strata of Sangiran (Djetis), but at a slightly higher level than that of the remains of *Pithecanthropus II* and *Meganthropus*, fragments of chalcedony and of objects similar to harpoons have been found; it is, however, doubtful that these remains have any connection with those of *Pithecanthropus*.

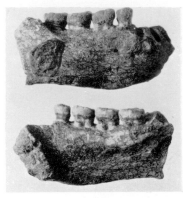

FIG. 16.8 *Pithecanthropus modjokertensis*, Sangiran mandible, fragment, external and internal view (by courtesy of G. H. R. von Koenigswald).

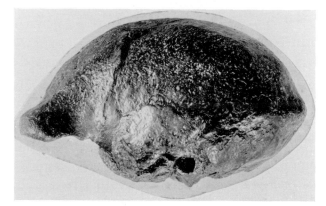

FIG. 16.9 *Pithecanthropus erectus* II, side view, Sangiran (by courtesy of G. H. R. von Koenigswald).

4. *Protoanthropus*

Sergi assigned the name "*Protoanthropus*" to that assemblage of forms which we consider to represent the first and most ancient types of Hominids. These forms, the most primitive we can attribute to man, are represented in Asia by the *Sinanthropus* (*Sinanthropus pekinensis*) of China, in Europe by the *Euranthropus* of Mauer (*Homo heidelbergensis*) and in Africa by the *Atlanthropus* (*Atlanthropus mauritanicus*) of Algeria.

Sinanthropus. The remains attributed to *Sinanthropus* (Fig. 16.10) come from fossiliferous deposits filling the cavities and anfractuosities of the limestone hills near the village of Choukoutien, about 50 km southwest of Peking.

The first find consisted of only three teeth which Davidson-Black, a Canadian anatomist in 1927 referred to as hominoid type called *Sinanthropus pekinensis*. An extensive campaign of excavations carried out between 1927 and 1930 led to a rich collection of material made up, as regards the hominoid fossils, of a large number of more or less fragmentary cranial bones (skull-caps, mandibles, fragments of maxillae, teeth) and a smaller quantity of

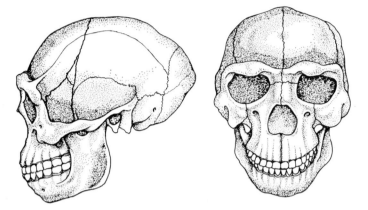

FIG. 16.10 Skull of adult female from Choukoutien according to the restoration of Weidenreich; lateral and frontal views.

fragments of long bones. It is a complex of remains which can be attributed to at least 40 individuals, about 15 of whom were juveniles. With them were collected many worked stones, still very crude, in spite of obvious signs of the use of fire (charcoal, ashes, smokey stones).

Dating of the remains with respect to basic chronology from the European glaciations has been made difficult by the fact that in China only two Pleistocene phases are clearly distinguished: an ancient and a recent. The bones of *Sinanthropus* certainly belong to the early Pleistocene but they are accompanied by a predominantly hot fauna, with some representatives of a humid climate, and they are found in a primary deposit slowly laid down.

The most likely opinion is that the remains of *Sinanthropus* go back to the end of the Mindel glaciation, that is, to the end of the middle Pleistocene, so that they would be about 200,000–250,000 years more recent than *Pithecanthropus*.

The morphological characteristics of *Sinanthropus* have been obtained, fundamentally, from six more or less incomplete crania, the left side of a maxilla, one half of an adult mandible and part of the chin of a child, seven fragments of femurs, two of humeri, one of a radius and more than 50 isolated teeth. In all the characters which can be studied there appears to be a strong sexual dimorphism.

The cranium of *Sinanthropus* has the same bursiform shape as that of *Pithecanthropus*, but the retro-orbital narrowing is somewhat less pronounced than in the latter. The dimensions of the cranium are a little larger than those of *Pithecanthropus*, particularly as regards width and height. The estimated capacity varies between 850 and 1220cc for the different crania, with an average value of 1000cc. Thus the capacity at its maximum equals that of some existing human populations (Andamanese, Vedda, New Guinean), and it should therefore be placed between that of *Pithecanthropus*, to which it is closer, and that of *Homo*. From the side the cranium is seen to be, on average, as high as that of *Pithecanthropus*. The supra-orbital torus is similar to that of *Pithecanthropus*, as is the occipital torus, but the latter does not continue in the supra-mastoid crest. The frontal bone, although still receding, tends to form an angle similar to that which is found in Neanderthal Man.

The facial skeleton has been hypothetically reconstructed on the basis of the morphology of a left maxilla. It exhibits a large, flat nasal region, low orbits, high zygomata and the absence of canine fossa.

Several fragments of the mandible are preserved, which reveal a tremendous variability in this bone: some mandibles resemble that of modern man, others are far removed from it. The curve of the dental arch and the conformation of the condyles, for example, are human in type; while the region of the symphysis is pithecoid. The teeth are larger than those of existing humans. The pulp cavities are very deep (taurodonts); the upper canine, which is rather strange, must have been extraordinarily strong and pointed, while the lower is spatulate as in man today.

Only a few fragments of long bones of the limb-skeleton are available. Notwithstanding some particulars of a monkey-like type (anterior-posterior flattening of the femoral diaphysis) they appear to be perfectly human and indicate that *Sinanthropus* was certainly endowed with an erect stance. It must have been of rather short stature: 150–160cm in the male.

Some workers have proposed changing the name of *Sinanthropus* to *Pithecanthropus erectus pekinensis* because of the great resemblance to, and to underline the close phylogenetic connection with *Pithecanthropus*.

There are, nevertheless, noteworthy differences, such as the cranial capacity and the clear separation existing between the occipital crest and the supra-mastoid crest, which place *Sinanthropus* on a higher level than *Pithecanthropus*.

Euranthropus. In 1907, a human mandible accompanied by very ancient-fauna was found in a deposit of Quaternary fluvial sands left by the ancient bed of the Neckar at Mauer, near Heidelberg (northern Baden), at a depth of 24m. Some of the remains (*Rhinoceros etruscus*) would suggest the Gunz-Mindel interglacial, others (*Elephas antiquus*) the Mindel-Riss. Most workers, basing their opinion on the morphological characters of the mandible as well, still prefer to attribute it to the latter period. In any case, it would appear to antedate *Sinanthropus*, and is certainly the oldest human type found to date in Europe.

The mandible is complete and all the teeth have been preserved except for the loss of the crowns of some of the premolars and of the first molars on the left. It is an extremely large mandible, with wide ascending rami which have small sigmoid grooves, rounded coronoid apophyses, and very large condylar articular surfaces. The body of the mandible is extremely thick; it lacks a chin, and all the lower part of the region of the symphysis is pushed backwards, with an anterior profile more pithecoid than human. Along with these pithecoid characters there are clearly human ones, such as the parabolic form of the alveolar arch, the absence of a diastema, the reduction of the canines (the crowns of which do not extend beyond the level of the other teeth), the reduction in volume of the third molar with respect to the other two, and the system of cusps. Taken altogether the mandible is not very far from those of *Sinanthropus* and *Pithecanthropus*, although it exhibits many human characteristics.

To underline better the importance of this fossil, whether in the geological sense or because of its morphological characteristics, it would perhaps be more suitable to call it *Paleanthropus* rather than *Euranthropus*, as was proposed by Bonarelli in 1909.

Atlanthropus. Represented by three mandibles, a parietal bone and a few teeth found at Ternifine, near Oran (Algeria), in 1954. The finds were accompanied by a fauna datable to the beginning of the middle Pleistocene (African Kamasian).

Two of the mandibles (Figs 16.11 and 16.12) are complete or almost so, and in general resemble those of *Sinanthropus* and *Euranthropus*, but also recall *Pithecanthropus*, *Telanthropus* and *Meganthropus* because of their robustness. Chronologically *Atlanthropus* has been placed at the end of the Mindel (400,000 years ago) and could be contemporaneous with *Sinanthropus* and perhaps a little later than the mandible of Mauer. The remains

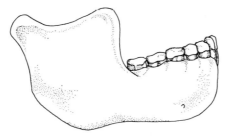

Fig. 16.11 Mauer mandible; right lateral view.

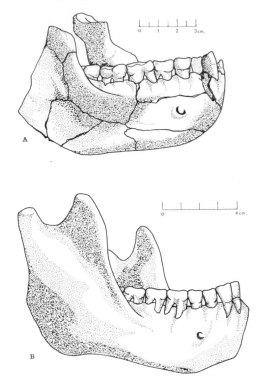

FIG. 16.12 Right lateral view of the mandible from Ternifine I (above), and Ternifine III (below).

were accompanied by a stone hand-axe industry of the Abbevillian-Acheulian with elements which may have been local variations.

Further finds of Pithecanthropomorphous teeth discovered at Rabat (Morocco) can probably be attributed to this group. The evolutionary continuity of this last group of finds (*Protoanthropus*) with that of *Pithecanthropus* is almost indisputable. They probably represent the expansion phase of that group of pre-human forms, whose most ancient representative would have been *Pithecanthropus*, which, in turn, would have ancestors in common with the Australopithecines.

This is not to say that some of the most recent forms of Australopithecines have not contributed, by a kind of introgressive hybridization, to the composition of the characters found in the mandible of *Atlanthropus*.

It should be worthwhile at this point to try to synthesize the differences in the characteristics of the skeleton and of the teeth which distinguish these two groups, and to try to individuate their evolutionary tendencies (Table 16.1).

TABLE 16.1. Differences in the structure of the skeleton and in the dentition, and evolutionary tendency revealed (von Koenigswald 1962).

Australopithecines	Pithecanthropus
Cranial bones thin	Cranial bones thick
Nasal aperture like that of the chimpanzee	Nasal aperture of human type
Superciliary arch (generally) slightly pronounced	Superciliary arch pronounced
Second upper molar larger than first	Second upper molar smaller than first (except for the most ancient types)
Lower posterior premolar with two roots	Lower posterior premolar with only one root
Dimensions of the brain (probably) increasing	Dimensions of the brain increasing strongly
Most recent rypes (Swartkrans, Olduvai) with medium crest	Even the most ancient forms (presumably) without median crest
Notable increase in size of molars and premolars	Reduction in size of molars and premolars
Canines and incisors considerably reduced	Canines and incisors less markedly reduced
Milk teeth molarized	Milk teeth remain primitive

5. Paleanthropus

In the desposits of the Riss-Würm interglacial and of the first phases of the Würm glacial all over the Old World (Europe, Asia and Africa), human types have been found that resemble each other in many characteristics and that can be described in certain detail because they are represented by numerous and fairly complete remains. Partly on the basis of morphological characters, but principally from their geographical distribution, humans of these types can be divided into: *Homo neanderthalensis*, (Europe); *Homo soloensis*, (Ngandong, Java; *Homo palestinus*, (Mount Carmel, Israel). We shall consider them separately.

Homo neanderthalensis (Europe). This group takes its name from Neanderthal, near Dusseldorf (Renania), where the skullcap of a cranium, which became famous for the discussions to which it gave rise, was found in 1856 together with the other bones of the skeleton.

The list of the finds of this type originating in this geological period in Europe is given in Table 16.2 and Fig. 16.13. In some places remains of this type of human being were found together with worked stone of the Mousterian industry. This fact suggests that Neanderthal men were the craftsmen

Fig. 16.13 Geographical distribution of the most important Neanderthal finds: 1, Neanderthal; 2, Spy; 3, La Chapelle-aux-Saints; 4, Le Moustier; 5, La Ferrassie; 6, La Quina; 7, Gibraltar; 8, Krapina; 9, Saccopastore; 10, Broken Hill (Rhodesian Man); 11, Saldanha; 12, Mount Carmel; 13, Shanida; 14, Teshik Tash; 15, Solo; and 16, Ma-Pa.

who produced stone implements of this kind. Other finds have furnished the information that they ate game and fish as well as plant foods, were canni-bals, knew the use of fire, were nomads, lived in caves and took refuge in rock shelters, buried their dead by placing them in special positions and burying with them both objects relevant to the deceased and votive offerings.

The general characteristics which can be ascribed to this group of finds are:

(a) a peculiar flatness of the brain case (platycephalia);
(b) a receding and flat forehead ending in the anterior part in a strong projection (supra orbital torus) which extends like a visor to limit the aperture of the orbit from above;
(c) a pronounced retro-orbital narrowing of the cranial vault united with a distinctive swelling of the inferior posterior region of the cranium.

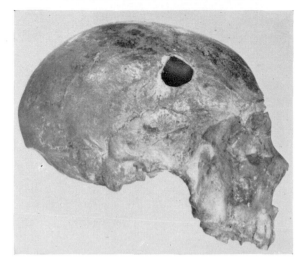

Fig. 16.14 Skull from Saccopastore I, lateral view (courtesy of the Instituto Italiano di Antropologia, Roma).

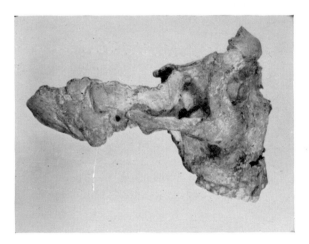

FIG. 16.15 Skull from Saccopastore II, lateral view (courtesy of the Instituto Italiano di Antropologia, Roma).

Seen from above, therefore, it assumes the shape of a purse (bursiform); (d) a very large face whether in absolute size or in relation to the braincase; (e) flat zygomatic bones directed somewhat obliquely forward, the jaws enlarged so that the appearance on the whole is of a sort of muzzle; (f) very large orbits;

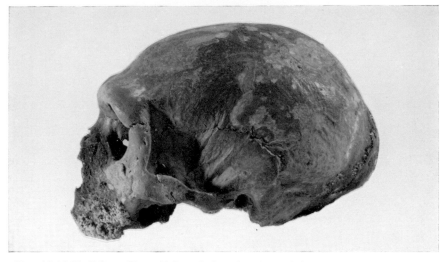

FIG. 16.16 Skull from Circeo I, lateral views (courtesy of the Museo Pigorini, Roma).

(g) extremely broad nasal aperture;

(h) prominent dorsum of the nose;

(i) very wide alveolar region;

(j) massive mandible with no chin or the bare beginning of one;

(k) large molars with deep pulp cavities (taurodontia), the premolars and canines not larger than those of present-day man;

(l) the spinous apophysis of the cervical vertebrae projecting backwards;

(m) narrow and deep thorax;

(n) narrow and high pelvis;

(o) the bones of the limbs typically human, although heavier and with some particular characters denoting primitiveness. If the footprints found in the Ligurian cave of Toirano, near Savona, really belong to a Neanderthal man, he must have had a foot similar to that of the present New Caledonians;

(p) medium or short stature (the stature of La Chapelle man is calculated to have been 161cm, that of the Neanderthals 163cm);

(q) the endocranial capacity reaches a value of about 1600cc in the La Chapelle cranium, a value sensibly superior to the average capacity of the European of the present time (1450cc). Other finds display a lower cranial capacity. In general the cranial capacity of Neanderthal man must have been 1400cc, with a variability falling between 1350 and 1735cc.

(r) the convolutions of the brain, although exhibiting general characteristics similar to those of modern man, are morphologically distinct because of their simplicity and coarseness.

Notwithstanding these common characteristics, from a morphological point of view it is possible to distinguish at least two forms: an earlier one living in the Riss-Würm interglacial (Gibraltar, Mount Maurin) and a later one living during the Würm glacial (Neanderthal, Circeo, La Chapelle).

The first form, also called "pre-Neanderthal" had a less capacious cranium than the second, and less specialized morphological characters (angle of the forehead, lack of occipital chignon, presence of canine fossae, natural flexion of the base of the cranium, mandible with rather archaic characters, receding chin, ascending ramus large, molars slightly taurodont).

The second appears to be highly specialized for adaptation to life in an extremely cold climate, very homogeneous and with the typical Neanderthal characters that we have listed.

The difference of degree of specialization is very important. The minor degree of specialization exhibited by the first form conferred on it a greater plasticity and possibility of adaptation. The high degree of specialization and the homogeneity of the second are perhaps the reasons for its lowered vitality, so that it became extinct without leaving descendants.

If Neanderthal man had had a genetic continuity, in fact, it would have to be sought in the more ancient branch. Before considering this problem, however we still have to describe other groups of remains; the African and Asiatic ones that, in contrast to the European groups living in glacial periods, may be called tropical types.

Homo rhodesiensis (Africa). In 1921, an almost complete cranium was found in the cave of Broken Hill, Northern Rhodesia, (Fig. 16.18) along with human bones whose attribution to the same individual or to individuals of the same type is still uncertain. The cranium displays very pronounced Neanderthaloid characteristics. The find appears to be relatively recent so that it has been assigned to the end of the last African pluvial (Gamblian), which corresponds to the last European glaciation; to an age, that is, of about 20,000 B.C. It is generally considered to be a "retarded" form which could survive in favourable environmental circumstances; its cranial capacity is about 1300cc.

Again in Africa, along the southwest coast, in Saldanha Bay, 190 miles north of Capetown, other remains have been found, represented by a skullcap and a small fragment of a mandible. The remains were accompanied by a very ancient fauna and by remains of an industry of the "Acheulian" type. On the basis of the associated faunal remains, the find must be attributed to the last interglacial (about 100,000 years ago), while in conformity with the accompanying industry, according to Coon, it must be of a much more recent time (40,000 years). In the first case, it would be contemporary with the Pre-Neanderthals, in the second with true Neanderthal man.

TABLE 16.2. Principal sites of Neanderthal finds in Europe

Sites	Date	Finds	Geological Age
Forbe's Quarry (Gibraltar)	1848	Incomplete female skull of adult	
Neanderthal (Dusseldorf)	1856	Male cranium and 13 bones of an adult skeleton	
La Naulette (Belgium)	1866	Incomplete mandible	
Sipka (Moravia)	1880	Fragments of skull and mandible	Würm glaciation
Spy 1⁰–2⁰ (Belgium)	1886	Two incomplete skeletons of adults	
Bañolas (Spain)	1887	Incomplete mandible	
Malarnaud (Ariège–France)	1889	Almost complete mandible of adolescent	
Krapina (Croatia)	1895––1905	Fragments from more than 20 skeletons of two sexes and of different age	Riss–Würm interglacial
Le Moustier (Dordogne)	1908	Skeleton of adolescent. Intentional sepulture	Würm glaciation
La Chapelle aux saints (Corrèze–France)	1908	Skeleton of adult male. Intentional sepulture	Würm glaciation
La Quina (Charente–France)	1908	Skeleton of adult female	Würm glaciation
La Quina (Charente–France)	1921	Skull of child. Fragments from different skeletons	Würm glaciation
La Ferrassie (Dordogne–France)	1909––1912	Fragments from six incomplete skeletons (two of adults). Intentional sepultures	
Jersey (Norman Isles)	1910	Thirteen teeth	
Ehringsdorf (Weimar)	1914––1916	Incomplete mandible of adult Infant mandible and the remainder of an infant skeleton	Riss–Würm interglacial

Location	Year	Description	Dating
Kiik-Koba (Crimea)	1924	Some bones of two skeletons without skulls	Riss–Würm interglacial
Ehringsdorf (Weimar)	1925	Fragmented cranium	Würm glaciation (first stage)
Gibraltar	1926	Parts of infant skull (5–6 years)	Riss–Würm interglacial or Würm glaciation
Gánovce (Slovakia)	1926	Parts of a skull with a natural outline in marble	Riss–Würm interglacial
Saccopastore I (Rome)	1929	Adult female skull without some frontal regions, including the zygomatic arch	Riss–Würm interglacial
Saccopastore II (Rome)	1935	Part of an adult male skull	Riss–Würm interglacial
Steinheim (Württenberg-Germany)	1933	Skull with a mutilated base	The last period of the second interglacial (Mindel-Riss) or Riss interglacial
Monte Circeo I (Rome)	1939	The skull of an elderly male	Würm glaciation
Monte Circeo II (Rome)	1939	Incomplete mandible	Würm glaciation
Montmaurin (France)	1940	Mandible	Riss–Würm interglacial
Monte Circeo III (Rome)	1950	Incomplete mandible	Würm glaciation
Arcy-sur-Cure (France)	1950	Alveolar parts of two maxillares with teeth	Würm glaciation
Játiva-Valenza (Spain)	1953	Right parietal bone	Würm glaciation

The industry associated with this find, among other things, would suggest that it is the most direct ancestor of man in Rhodesia.

Homo soloensis (Java). In the island of Java, in the fossiliferous layer of Ngandong, near the Solo river, a number of attempts (1932, 1937, 1939) brought to light eleven crania, some of them badly damaged, and two

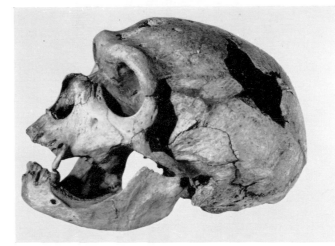

FIG. 16.17 Skull of man from La Chapelle-aux-Saints. (courtesy of Musée de l'Homme Paris).

tibiae. The stratum corresponds to a rather arid period which could be attributed to the interglacial Würm II. The cranial capacity of this find is certainly over 1000cc. The forehead is narrower and more receding than that of the European Neanderthalers; it exhibits a pronounced chignon and large mastoid processes. Because of many of these characters and because of their antiquity the remains of Solo Man can be considered to be those of the most primitive of the Neanderthal men; something intermediate between the latter and *Pithecanthropus*.

Siberian Man. This type is represented by a skeleton of a child found in 1940 in the cave of Teshik-Tash (Uzbekistan) (Fig. 16.19). The skeleton was found surrounded by goat horns arranged in an orderly fashion in pairs, which suggests an intentional burial. It displays characters which place it close to the European Neanderthal man.

Other finds of *Paleanthropus*. Remains of the Paleanthropic type other than those described above have been found: Crimean Man represented by

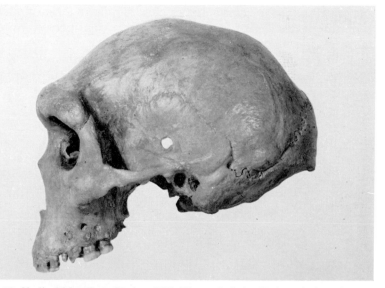

FIG. 16.18 Skull of Man from Broken Hill (*Homo rhodesiensis*), lateral view. (courtesy of British Museum, London).

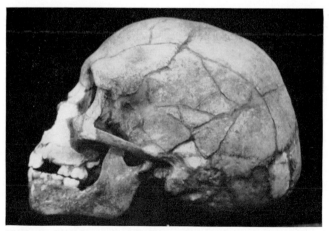

FIG. 16.19 Skull of child from Teshik Tash (Uzbekistan, Central Asia); lateral view (courtesy of the Anthropology Institute of Moscow).

skeletal remains of an adult and a child. It has typically neanderthaloid characteristics.

Let us now return to the Neanderthal remains of Europe and North Africa. They can be roughly divided into two groups: one group tending to specialize in the direction of typical Neanderthal characteristics and the other tending to develop characteristics more similar to those of man as he exists, so that it has come to be considered as a transitional form on the way to *Homo sapiens*.

To the first group can be assigned the remains of Ehringsdorf, Krapina, Gànovce and Saccopastore, as well as those of Gibraltar, Sidi-Abderrahman

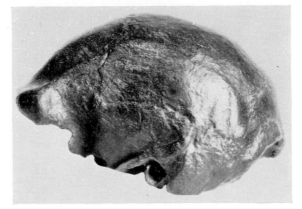

FIG. 16.20 *Homo soloensis*, Upper Pleistocene, Ngandong (by courtesy of G. H. R. von Koenigswald).

and Temara in Morocco, and that of Haua Fteah in Cyrenaica. The second group, which is more interesting to us because it exhibits the characteristics of modern man, can be introduced with the most ancient of them: the Steinheim cranium. This cranium was found in the place of the same name, in Wurtemberg, in 1933. The cranium is complete, although the mandible is missing. It must have belonged to a female. It can be dated to the Mindel-Riss interglacial with a high degree of probability.

Although it presents a noticeable supra orbital arch of the same type as that of *Sinanthropus*, the cranial vault is very high. Its cranial capacity must be comprised between 1100 and 1200cc. The molars are small. The orbits are widely spaced; the root of the nose is sunken as in existing human beings. The attachment of the zygoma differs from that of the other Neanderthals.

Because of this intermingling of diverse characters the Steinheim cranium represents an important point in human evolution. If the Neanderthal characters are underemphasized it could easily be classified among the ancient remains of *Homo sapiens*, if instead these traits are overstressed it is possible to classify it among the typical Neanderthals. To this human type is linked all that group of finds which Sergi indicates by the name *Prophaneranthropi*. He considered them to be the precursors of the hominids that developed in the upper Palaeolithic and which, because they were endowed with a morphology similar to that of present-day man, he called *Phaneranthropi*.

The most important representatives of this group of fossil hominids are Swanscombe man, Fontechevade man, the mandibles of Montmaurin, Olmo man, and Quinzano man. The Swanscombe finds were discovered in a cave near Swanscombe, Kent (England). The find consists of the occipital bone and two parietals. The cranium has a capacity of 1325cc and is rather low and wide. The endocranial cast does not reveal any significant difference with respect to the cranium of a Neanderthaler; the measurements, however, make it almost the same as that of a living man, while distinguishing it clearly from that of Neanderthal man.

The find is certainly very ancient, probably antecedent to the entire Riss and thus contemporary with the most ancient Neanderthals. It was found associated with a hot weather fauna and an Acheulian industry.

Fontechevade man is represented by the remains of the crania of two individuals. The stratum from which the two fragments come has been dated to the Riss-Würm interglacial; therefore this find could be more or less contemporary with the most ancient Neanderthals.

The find from Olmo, a small village near Arezzo in Italy, lacks a face, although it comprises almost all the forehead and part of the parietals and the occipital bone. It has a capacity of almost 1400cc. The assemblage of morphological characters, especially the absence of a supraorbital torus, and the upright, even if low, forehead, make it much more similar to modern man than to the Neanderthalers. It was found together with flakes of the Mousterian type. Notwithstanding the existence of doubts on the exact stratigraphy, it seems that the find is contemporary with at least the last Neanderthals.

The find from Quinzano, a small town in the environs of Verona, is represented by a piece of the occipital bone, which in form, thickness and dimensions resembles the finds of Swanscombe and Fontechevade.

Another interesting find is the mandible from Montmaurin in the Haute-Garonne which according to Vallois comes from the Mindel-Riss interglacial.

As well as these finds in Europe, there is another important series of skeletons in Palestine. They demonstrate indirectly the existence of primitive *sapiens* in the Middle East, a mixed group of clearly neanderthaloid types. They come chiefly from the caves of Mount Carmel in the vicinity of Haifa and Galilee (Fig. 16.21). A large number of representative finds were discovered at this site. Many of them were found in a stratum characterized by a hot-weather fauna and with an industry considered to be like the Mousterian (Levalloisian-Mousterian). The finds have been placed in the first phases of the sea regression corresponding to the marine recession at

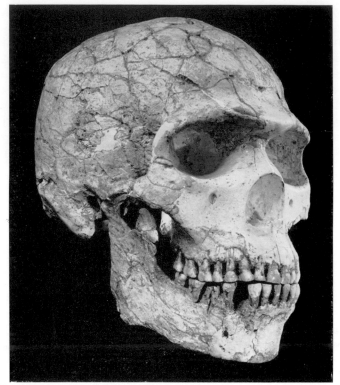

FɪG. 16.21 Skull from Skhul V, after reconstruction; view of the profile. (Courtesy of the Peabody Museum, Harvard University).

the beginning of the Würm, that is about 60,000 B.C.

This assemblage of remains is remarkable mainly for the wide variability it exhibits; and for the association in different individuals of Neanderthal and phaneranthropic characters. Thus, along with the supraorbital torus, the platycephalia, the complete prognathism, the absence or reduction of the canine fossae and of the chin, it is possible to observe a more rounded forehead, the tendency of the supraorbital torus to be divided, the presence of canine fossae, a well-modelled chin, and above all a cranial vault notice-ably higher and more rounded. The characteristics of this group of finds are therefore extremely variable—more variable than those of the individuals of any known population.

Two hypotheses may be called upon to explain this great variability: either the neanderthaloid type had become unstable and was being trans-formed into *sapiens*, or the remains from Mount Carmel are a mixture of two types. Both possibilities have been considered by various experts. However, the second seems more probable, especially if the type of culture

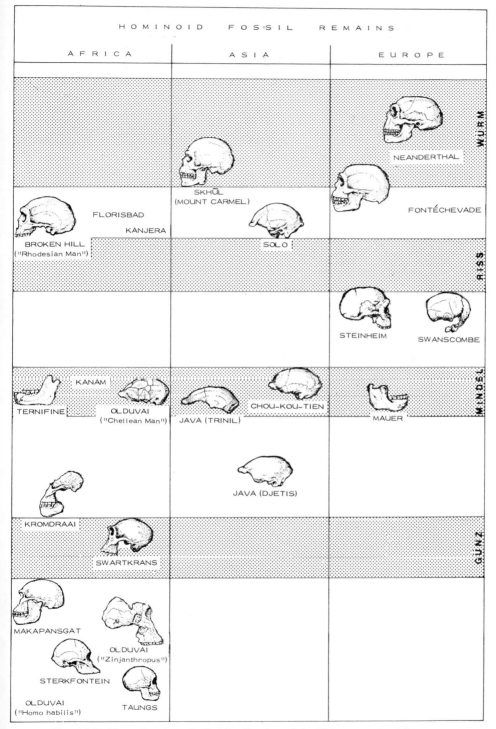

FIG. 16.22 Tentative chronological localization of the Pleistocene Hominids.

with which these remains were associated is considered. This culture, called pre-Aurignacian-Mousterian is, in fact, according to Rust (1958), a hybrid culture composed of Neanderthal elements mixed with elements of the *sapiens* type. The finding of remains having such variable characters is, however, an important indication of the genetic continuity of Neanderthal man with modern man.

6. *Phaneranthropus*

In Europe, as in Asia and Africa, a profound change in human types corresponds with the upper Palaeolithic. The *Palaeanthropi* (Neanderthal man) became extinct and human types appeared with characteristics different from those of the palaeanthropic types of the Middle Palaeolithic. These new forms are certainly connected with the Neolithic forms and with modern man; consequently we can ascribe to them the name *Homo sapiens fossilis*. This aspect, however, is beyond the scope of this synthesis.

GENERAL REFERENCES

Boule, M. and Vallois, H. (1952). "Les Hommes fossiles", Masson et Cie, Paris.

Brace, L. (1967). "The Stages of Human Evolution", Prentice-Hall, Englewood Cliffs, N.J.

Brace, L. and Montagu, M. F. A. (1965). "Man's Evolution. An Introduction to Physical Anthropology", The MacMillan Co., New York.

Chavaillon, J. (1970). "Découverte d'un niveau oldowayen dans la basse vallée de Omo (Ethiopie)". *Bull. Soc. Préhist. Franc.* **67**, 7–11.

Coon, C. S. (1962). "The Origin of Races", Alfred Knopf, New York.

Day, M. (1965). "Guide to Fossil Man", Cassel and Co., London.

Jullien, R. (1965). "Les Hommes Fossiles de la Pierre Taillée", Boubée and Cie, Paris.

Korn, N. and Smith, H. R. (eds) (1959). "Human Evolution: Readings in Physical Anthropology", Holt, Rinehart and Winston, New York.

Lasker, G. W. (ed.) (1960). "The Processes of Ongoing Human Evolution", Wayne State University Press, Detroit.

Robinson, J. T. (1968). The origin and adaptive radiation of the Australopithecines. *In* Sanderduuck aus "Evolution und Hominisation", (G. Kurth, ed.), J. Fisher Verlag, Stuttgart.

Sergi, S. (1962). The Neanderthal Palaeanthropi in Italy. *In* "Ideas of Human Evolution" (W. Howells, ed.) Harvard Univ. Press, Cambridge, Mass.

Sergi, S. (1967). I tipi umani piu antichi Prehominidi e Ominidi fossili, *In* "Razze e Popoli della Terra", (R. Biasutti ed.), Vol. 1, pp. 184–162, UTET, Turin.

Tobias, P. V. (1968). The taxonomy and phylogeny of the australopithecines *in* "Taxonomy and Phylogeny of Old World Primates with References to the Origin of Man", B. Chiarelli (ed.), Rosenberg and Sellier, Turin.

von Koenigswald, G. H. R. (1962). "The Evolution of Man", Univ. of Michigan Press, Ann Arbor.

Washburn, S. L. (ed.) (1964). "Classification and Human Evolution", Methuen and Co., London.

17

Posture of the Primates and the Acquisition of an Upright Posture by Man

1. THE POSTURE OF THE PRIMATES

The posture of the living primates varies according to the conditions of life to which they are adapted.

The baboons (*Papio*) live in savannah country, rocky regions or places with scattered vegetation; they move on all four feet. Their gait is quadrupedal, even though they are capable of using all four extremities as hands.

The Cercopithecines live for the most part in trees and use all four limbs for climbing and maintaining their equilibrium on the branches. Their gait is quadrumanous, but they consistently use their lower limbs for grasping and their arms for procuring food.

The gibbons (*Hylobates*) climb trees using their upper limbs (which are noticeably more developed than the lower ones) almost exclusively. The lower limbs are used for movement in a horizontal direction, like an acrobat, along the branches. Consequently, the gibbon is defined as a brachiator with a predominantly vertical posture.

The anthropoid apes show great differences among themselves.

The orangutan has adaptations best suited to an arboreal existence and consequently ones that somewhat resemble the adaptations of the gibbon; the gorilla and the chimpanzee are better adapted to a terrestrial life and thus have adopted features similar to those of the baboon.

Man is bipedal with an erect posture; and hands that are completely free to grasp objects.

Obviously these different conditions will be accompanied by noticeable structural differences in the skeletal and muscular apparatus and the organs of equilibrium.

If it is accepted that these divergent types of posture are a result of the adaptation to different conditions of life during the evolution of the various primate species, the problem, then, is to determine the posture of the first form of primate.

It is highly probable that the first primates lived for the most part in trees; they must therefore have been good climbers. Confirmation of this can be seen in the existing prosimians (the direct descendants of the first primates) who lead an arboreal life of this type. Furthermore, it was the only possible way of life for nearly defenceless animals like the primates in an epoch during which the earth's inhabitants had to suffer the violence of the first carnivores and the competition of the first rodents and lagomorphs.

The terrestrial adaptation of the baboons, some macaques, the gorilla and, to a lesser degree, the chimpanzee must therefore be considered as a later event. This radical change was probably due to the retreat of the forests and the consequent necessity of conquering new habitats, or to their bodies becoming so bulky that an arboreal life was difficult and exhausting.

In this regard it is significant that the female and young gorillas (who are less heavy) arrange their nests for the night in the branches of trees, while the males, as a rule massive and ponderous, arrange theirs on the ground.

What posture did the first hominids assume? Was an erect stance an achievement which resulted from being a brachiator like the gibbons, or from having a semi-erect carriage like that of the chimpanzee? For reasons which will be given, it is assumed that it was an achievement secondary to a quadrupedal condition of the same type as that of the African apes (gorilla, chimpanzee). Recent research, especially that conducted by Kortlandt, has demonstrated that the chimpanzee can walk upright on its lower limbs for long periods when it is obliged to do so.

Among these experiments, two are especially significant. In the first instance Kortlandt, placed a group of chimpanzees on an isolated piece of ground and, not far away, put a stuffed leopard. The chimpanzees having vainly attempted to flee, arranged themselves into a group, with the males and older females in front, the youngsters and mothers behind, and organized a charge against the leopard. Step by step as they advanced toward their enemy the chimpanzees gathered up sticks, stones and every possible object

FIG. 17.1 Different types of posture in Primates; (A) *Tupaia*, a primitive type of posture in Primates; (B) *Saguinus*, arboreal adaption; (C) *Hylobates*, arboreal specialization (brachiator); (D) *Macaca*, terrestrial specialization; (E) *Pan*, semiterrestrial adaption; (F) *Hominid* incipient erect posture; (G) *Homo*, erect deambulation (redrawn from different sources).

they could find on the ground. Their quadrumanous carriage became semi-erect and in this attitude they prepared for the final attack (Fig. 17.2).

On another occasion Kortlandt placed a young chimpanzee in an environment in which living conditions generally were extremely difficult, and food was hard to find. One day the chimpanzee was permitted to discover twenty-odd coconuts, bananas and other dainties. Wishing to carry this food-stuff to a secret hiding-place, the chimpanzee filled first his mouth, then his hands and then finally his lower extremities. However, burdened in this way he was unable to move. Suddenly, as if struck by a brilliant idea, he folded one of his upper limbs and placed it against his body. He arranged the coconuts between this arm and his chest, took the rest of the fruit in his other hand, and then, semi-erect on his hind legs, he ran towards his hiding-place.

2. THE ACHIEVEMENT OF UPRIGHT POSTURE AS DEDUCED FROM THE EXAMINATION OF AUSTRALOPITHECINE SKELETONS

Although the chimpanzee occasionally assumes an erect stance, in men it is normal. How has this come about?

Among the most ancient representatives known of the Hominid family are numbered the Australopithecines. Of these, as we have seen, there exists a large number of remains, even of the long bones. Let us see what data we have for establishing their type of posture.

First of all, the five or six places in which the remains of these hominids have been found were not caves, but more or less solitary places where they probably stayed only occasionally, and where they caught and ate their prey. This is shown by the large quantity of animal remains found with them. At Makapansgat for example, 92% of the remains were made up of antelope bones of various sizes and ages, 1·27% of baboon bones and only 0·26% of Australopithecine remains.

Moreover, Dart has observed that a very high percentage of the baboon remains found have a depression on the cranium which suggests that it was caused by a blunt instrument, probably a large ungulate femur. It is therefore very probable that the Australopithecines were hunters.

Firm data which show that the Australopithecines used objects of bone, wood or stone do not exist. There is, however, proof that the carrying of stones took place. Such an activity, like the hunting mentioned above, can be carried on by a hominid only if its hands are free, and only if it can walk and run.

A stable upright carriage and adaptation for running cause many changes in the post-cranial skeleton which should be apparent even in fossil finds.

(a)

(b)

FIG. 17.2 (a) Male chimpanzee charging with a club at a leopard in a perfect erect posture and (b) a female with child swinging a 6 foot club immediately before delivering a hard blow on the stuffed leopard.

Fortunately, the remains of such parts of the skeleton of the Australopithecines are numerous, and can therefore be studied in detail.

The number and morphology of the lumbar vertebrae are significant in this regard. There are six found in the Australopithecine remains, while in

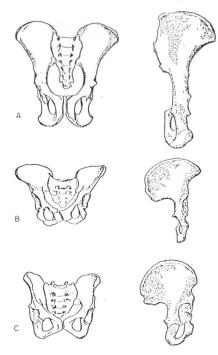

FIG. 17.3 Comparison of the shape of the pelvic bones in (A) the chimpanzee (B) *Australopithecus*, and (C) modern man (redrawn and combined from Le Gros Clark, 1964 and S. Washburn, 1969).

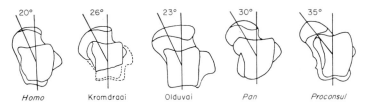

FIG. 17.4 The astragalus of the Australopithecines and some other Primates: (A) *Homo sapiens*, (B) Kromdraii, (C) Olduvai youth, (D) baboon, (E) chimpanzee, and (F) *Proconsul nyanzae*. The angle of the neck of the astragalus in the Australopithecines is intermediate between that of Man and the anthropoid apes and baboon. (After Coon, 1962 redrawn.)

man there are usually five and occasionally six or four. In the chimpanzee four. There are in the gorilla three or four. Their morphology makes it possible to distinguish the typically human curvature of the vertebral column.

The pelvis is a particularly important structure for upright posture, since it is the bony scaffolding on which the whole body rests. Four more or less complete Australopithecine pelvises have been found, two apparently from youths and two from adults.

Comparing the pelvic bones of the Australopithecines with those of the chimpanzee and of man, it is clear that all four finds resemble those of man in general form and in many details (as for example the anterior inferior iliac spine), but differ from them in some particulars (Fig. 17.3). The differences consist principally in the absence of body relief for the attachment of the gluteal muscles on the outer surface of the ilium, and a remarkable variability in the size of the ischial tuberosity.

These data have made it possible for various authors to put forth the hypothesis that these hominids had either a semi-erect posture or else an erect one for at least part of the time.

However, if the Australopithecines succeeded in hunting, and in carrying the spoils of the hunt, they must have had a carriage that on the whole was upright, and they must have been able to run. In order to run with the trunk erect, it is necessary to have a specially adapted lower limb.

Unfortunately only one pair of proximal bones and the distal end of a femur are available for study. From the proximal parts it can be seen that although the area of the trochanters has almost the same conformation as in humans, the muscle attachments are not as well developed. The distal part is rather small, but clearly human in type and sufficiently robust to support the entire weight of the body, not only for part of the time but constantly. Neither a tibia nor fibula has been found thus far.

Among the bones of the foot the talus is of great importance because it is interposed between the leg and the foot proper, and thus it can give indications of the rigidity or mobility of the plantar arch. The talus of the Australopithecines can be compared with those of man and other primates.

The neck-axis angle is intermediate between the angle found in the chimpanzee and that of man, but closer to the human one (Fig. 17.4).

Hunting requires not only an upright stance (and thus free hands), but also upper limbs that are especially agile and mobile.

In the Australopithecine remains the shoulder girdle is represented only by fragments of one clavicle and one scapula. The scapula has a long spine nearly human in form. The acromion projects as in man and in the anthropoid apes, but, unlike the latter, it is thin and flat (as in man). The area around the glenoid cavity conforms to that of the great apes, and is naturally robust, which denotes a remarkable muscular strength (Fig. 17.5).

The upper limb is represented only by fragments of the humerus and of some phalanges. The fragments of the humerus suggest a length of about 30cm and its structure is intermediate between human and anthropoid. The phalanges show signs of great agility.

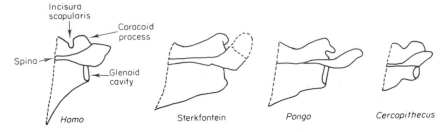

FIG. 17.5 The scapula of *Australopithecus*, Man, and other Primates.

In general, therefore, the post-cranial skeleton of the Australopithecines displays an aspect intermediate between that of human beings and that of the anthropoid apes, but it is closer to the human type. An examination of the human skeleton makes it clear that this being had an erect carriage and walked on two feet almost like modern man.

3. Relative Position and Orientation of the Foramen Magnum in the Process of Humanization

Other indications of the erect posture of the Australopithecines are provided by the position and the orientation of the foramen magnum. In the series of mammals it tends to move from a posterior position (dog) to an inferior position (man). This translocation is gradual in the primates. In the hominids the process is effected simultaneously with the acquisition of an upright carriage and the enlargement of the braincase.

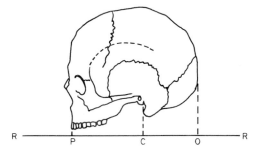

FIG. 17.6 Valuation of the position of the occipital forum. RR=section of the horizontal plane; P=projection of the anterior point (prostion); C=projection of the condilar point; O=projection of the posterior point (opisthocranium).

The position of the foramen magnum can be evaluated by means of an index proposed by Zuckerman. The cranium is oriented according to the plane of Frankfourt; the maximum length of the cranium and the face together are traced; and the point midway between the occipital condyles is projected onto this segment. The relationship between the posterior and anterior parts of this length gives the value of the index. Thus if P is the anterior point (generally represented by the horizontal projection of the prosthion), O the posterior point (generally represented, in man, by the horizontal projection of the opisthocranium) and C is the projection of the condyles (condylar axis), the index is measured by the relationship $OC:PC$ multiplied by 100. Zuckerman finds the value of this index to be always over 60 in man, equal to or less than 30 in the anthropoid apes, and 40 in the Australopithecines (Fig. 17.6).

The significance of the different intensities shown by this index can also be expressed as a relationship between several numbers:

Man	$OC:PC = 3:5$
Anthropoid Apes	$OC:PC = 1{\cdot}5:5$
Australopithecines	$OC:PC = 2:5$

In this way it can be seen that the distance from the condylar axis to the posterior point of the cranium (occiput) in the apes is about half of that distance in man, and that the Australopithecines occupy an intermediate position somewhat closer to the apes.

It thus seems evident that the Australopithecines, while they had a semi-erect or erect posture of the human type, had not yet reached the stage of holding the head completely erect. However, they represent a phase in the evolutionary process which led to the erect carriage of human beings, which certainly originated in a posture of the terrestial quadrumanous kind which can be seen today in African apes.

This was one of the first and one of the most important steps towards humanization. The stimulating agent, very probably, was the necessity of finding food in an environment complicated by the competition of other mammals, particularly the carnivores.

We shall now observe the most important variations which the various parts of the skeleton underwent during this complex process.

4. THE SHIFT OF THE VERTEBRAL BEAM TO A VERTICAL POSITION

Although the term "vertebral column" is used generally, it has a meaning related to its function only if the column referred to is in a vertical position (as in man). To be correct the term "vertebral beam" should be used for the quadrupeds, since their axial skeleton has only that function from a mechanical point of view. The axis of the vertebral beam of the quadrupeds, in normal conditions, lies parallel to the ground and so perpendicular to the lines of force of the earth's gravitational field. It is, therefore, subjected to

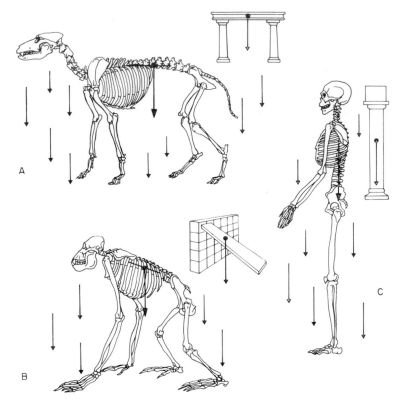

FIG. 17.7 The "vertebral beam" and the "vertebral column": indicating the forces to which they are subjected (A) dog, (B) Chimpanzee, (C) Man. (Courtesy of M. Masali.)

forces of flexion and bending even under static conditions. The beam is noticeably rigid, in that, the cohesion of its elements (vertebrae) is determined by the compression exerted on the bodies in a longitudinal direction by the muscles. Consequently, the above mentioned forces are discharged more or less uniformly onto the anterior and posterior limbs according to the particular postural architecture of each species (Fig. 17.7).

The situation is somewhat different in the clinograde primates, (those having a more or less oblique posture), such as the anthropoid apes. In this group the support, and therefore the discharge of forces, falls mostly on the posterior section, while the anterior part has only the function of secondary support. From the mechanical point of view the rachis is still a beam, but part of the gravitational forces are discharged longitudinally because of the effect of the inclined planes which make up the faces of the vertebral bodies. In man, the situation is very different, because the support of the

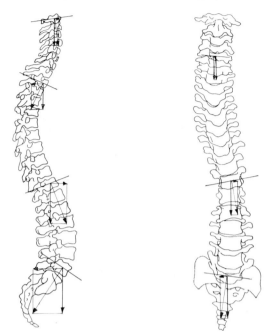

FIG. 17.8 Statics of the human vertebral column in relation to the systems of curvature (sagittal and lateral). (Courtesy of M. Masali.)

anterior limb (which thus becomes superior) is missing, and so the discharge of all the forces must, by definition, take place along the longitudinal axis of the spine. It is thus a true column, and has as its base the superior face of the sacrum and as its capital the atlas, which supports the human architrave, the head.

In this case not only the forces determined by the pressure exercised by the muscles have an effect on the vertebral bodies, but also those of gravity and probably those of longitudinal inertia (as in a descending leap). Consequently, if the vertebral column were a rigid and geometrically perfectly columnar body, it would have to have dimensions, cohesion, and resistance to compression notably greater than that which it actually possesses.

The system of overlapping curves, sagittal and lateral, give to the spine a helicoidal structure, allow it to react in an elastic manner to the demands made on it, and to discharge part of the kinetic energy horizontally. In this way the energy is dissipated through the muscular structures, bones and tendons of the trunk, and the curves at least ensure that all the pressure is not strictly perpendicular to the vertebral bodies (Figs 17.8 and 17.9). Important variations in the dimensions and in the morphology of the vertebrae follow from this.

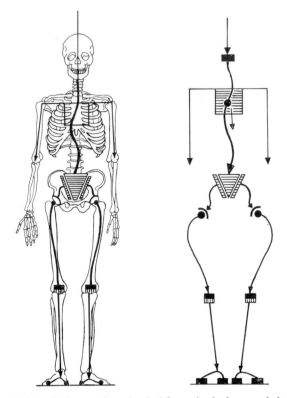

Fɪɢ. 17.9 Model of the discharge of mechanical forces in the human skeleton. (Courtesy of M. Masali.)

Particularly evident are the modifications in the size and shape of the vertebrae which follow the acquisition of an erect carriage. In the quadrupeds each vertebral body must support, under static conditions, only the forces of compression necessary to maintain the rigidity of the spine. However, with upright posture each vertebra must also support the burden of the masses which weigh on the underlying vertebrae. From the cervical to the lumbar vertebrae there is, thus, a progressive increase in the load and consequently a gradual increase in the dimensions of each successive vertebra. Thus in humans the "vertebral column" has the appearance of a cone-shaped trunk, which does not appear, or at least is not necessary, in the quadrupeds and is less evident in the clinogrades.

The most obvious adaptive manifestation in the vertebrae to the upright position is a shortening of the vertebral bodies in a head-to-tail direction and an increase in their transverse diameters. The vertebrae of the quadruped monkeys are in fact particularly elongated in the lumbar region;

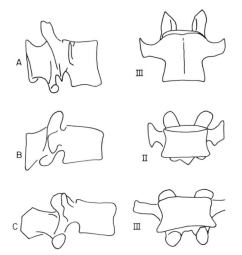

FIG. 17.10 The second and third lumbar vertebrae of Primates with different postures (A) *Cercopithecus*, (B) *Gorilla* and (C) *Homo*.

in the great apes the height of the body tends to be almost equal to its transverse diameter; while in man the bodies are predominantly wider than they are high.

The sacral region also undergoes interesting adaptations to an erect posture. A main function of the sacrum is to discharge to the pelvis the mechanical stresses to which the spine is subjected. In quadrupeds the sacrum is placed below the medial margins of the ilium and thus it functions as a support to the weight of the spine and of the abdominal viscera, which are directly or indirectly attached to it. The weight of the thoracic region on the other hand is discharged on to the shoulder girdle. Consequently, the sacral bone is generally thin, elongated and little differentiated from the rest of the spine. In the monkeys, on the contrary, it is much longer and more like a wedge between the ilia; in man it appears typically wedge-shaped and assumes its full function.

In an erect stance the purpose of the sacrum is to transmit to the hip bone all the stresses which burden the vertebral column (the weight of the trunk, the upper limbs, and the head; the traction of the muscles of the lumbar region; and, as well, the load of forces and occasional inertial demands involving the upper part of the trunk). Because of the special structure of the sacrum, the forces F, transmitted vertically through the spine, are divided into two resultants: oblique F' and vertical F'' of the primary force F, as the effect of the inclined planes constituting the sacro-iliac articulating surfaces. This mechanism permits the uniform distribution of the load to the lower limbs (Fig. 17.11).

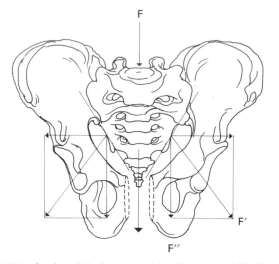

FIG. 17.11 Statics of the human pelvis. (Courtesy of M. Masali.)

5. Process of Development of a Brachymorphous Pelvis

Along with the adaptation of the sacrum and the vertebral column to the new conditions, the morphology of the pelvis assumes a new guise directly related to the new carriage. In the quadrupeds the abdominal viscera are suspended from the pelvis, while in man they rest in a "basin" formed by the pelvic bones themselves. Thus the term "basin" in reference to the pelvis is correct only in the context of orthograde beings.

The pelvis of the quadrupeds functions not only in locomotion, but also, as has been seen, in sustaining and discharging to the limbs the weight of the posterior part of the body. It is helped in this by the static function of the shoulder girdle. In man, the pelvis assumes a fundamental rôle in the statics and dynamics of erect posture, especially because it provides for the insertion of the gluteus muscles, which, in man, assume a particular configuration. They are very wide and the width is superior to the length relative to that found in the clinograde and quadrupeds.

The function of the glutei (which extend to cover the ischial tuberosities), while almost exclusively locomotor in the chimpanzee (Sperino), in man work to keep the trunk upright. Enormous muscular force is required to achieve this, and a balance of muscle tone must be obtained otherwise the erect carriage would be too tiring. For this purpose wide and comparatively short muscles are the most efficient; and to correspond with the muscular morphology an increase in the width of the iliac bones, and a simultaneous reduction in their length are also seen.

An erect stance is therefore accompanied by a large, wide and low (brachy-morphous) pelvic basin, characteristic of man and of the fossil hominids (Australopithecines). This is in direct contrast to that of the quadrupedal and arboreal monkeys (including the anthropoid apes), which is narrow and long (dolichomorphous). Brachypelvia (wide, low basin) is therefore a fundamental criteria for erect carriage in fossil finds.

GENERAL REFERENCES

Clark, W. E. Le Gros (1964). "The Fossil Evidence for Human Evolution". The University of Chicago Press, Chicago.

Coon, C. S. (1962). "The Origin of Races". Alfred A. Knopf, New York.

Delattre, M. A. (1958). "La formation du crâne humain". In "Les Processus de l'Hominisation". Editions du Centre National de la Récherche Scientifique, Paris.

Delattre, A. and Fenart, R. (1960). "L'Hominisation du Crâne". Editions du Centre National de la Récherche Scientifique, Paris.

Delmas, M. A. (1958). "L'acquisition de la station erigee". In "Les Processus de l'Hominisation". Editions du Centre National de la Récherche Scientifique, Paris.

Kortlandt, A. (1967). Experimentation with chimpanzees in the wild. "Progress in Primatology" (Stark, ed.) pp. 208–224.

Kortlandt, A. and Kooij, M. (1963). "Protohominid Behaviour in Primates (Preliminary Communications). In "The Primates" (Symposia of the Zoological Society of London, 10), (J. Napier and N. A. Barnicot, eds.), pp. 61–68, Academic Press, London and New York.

Masali, M. (1968). The ear bones and the vertebral column as indication of taxonomic and postural distinction among Old World Primates. In "Taxonomy and Phylogeny of Old World Primates" (B. Chiarelli, ed.), Rosenberg and Sellier, Turin, Italy.

Napier, J. R. and A. C. Walker (1967). Vertical clinging and leaping a newly recognized category of locomotor behaviour of Primates. Folia Primatologica 6, 204–219.

Preuschoft, H. (1971). Body posture and mode of locomotion in early Pleistocene Hominids. Folia Primatologica 14, 209–240.

Sigmon, B. A. (1971). Bipedal behavior and the emergence of erect posture in Man. Am. J. Phys. Anthrop. 34, 55–60.

Sperino, G. (1897). "Anatomia del Scimpanzé in rapporto con quella degli altri antropoidi e dell'Uomo". Un. Tip. Torinese (UTET), Turin.

Tuttle, R. H. (1969). Knuckle-walking and the problem of human origins. Science N.V. 166, 953–961.

Washburn, S. L. (ed.) (1964). "Classification and Human Evolution" Methuen and Co., London.

18

Differentiation of the Upper and Inferior Limb in the Primates and the Opposability of the Thumb in Man

1. DIFFERENTIATION OF THE LIMBS IN RELATION TO THE TIME DURING WHICH BRACHIATION HAS BEEN PRACTISED

Considering the analogous function of the anterior and posterior extremities in quadrupedal locomotion, we shall now see how, from the basic limb design of the four-footed animals, the limbs of the primates became specialized in relation to the different locomotor adaptations realized in this taxonomic group. All the species of primates are characterized by some sort of locomotor specialization, and even when a quadrupedal condition is manifested, it is always imperfect.

In an arboreal environment the differentiation of the limbs is linked to the techniques used in climbing, and to the weight of the body. Obviously a small monkey will have less need for differentiation of the limbs than a heavy anthropoid. While quadrupedal locomotion on tree branches might be possible for the former, it would certainly be problematical for a gorilla. In fact brachiation by suspension, with or without the help of the lower limbs, is used frequently by the largest monkeys. During brachiation the weight of the animal is suspended by the anterior limbs, while the posterior ones function as a support and to give impetus in a leap. From this stems the notable differentiation in robustness or in arm length seen in the great brachiators (orangutan, gibbon) in which the upper limb is much longer than the lower; while in those that were brachiators for a shorter period (gorilla, chimpanzee) the difference is less accentuated.

In the small arboreal monkeys with a prehensile tail (Cebidae) the posterior limb is slightly longer. In the terrestrial quadrupedal monkeys (baboons), the anterior and posterior limbs are equal.

In man (time of brachiation nil) an inverse kind of differentiation appears. Support and locomotion are entrusted exclusively to the lower limb, while the upper functions only in grasping. There is, therefore, a reversal of the limb proportions, so that the lower limb is more robust and longer than the upper.

Mobility is also differentiated; while the freedom of movement of the lower limb is limited in practice to the sagittal plane alone (mechanically only one "degree of freedom"), the upper extends its field of action to the three basic dimensions in space (three degrees of freedom). The strictly arboreal adaptations of the anthropoids and the terrestrial specializations of man presuppose distinct patterns of evolution. Therefore it seems evident that the "common ancestor" must have been undifferentiated from the point of view of locomotion.

2. Adaptation of the Lower Extremity for Upright Locomotion

The modifications which the foot have undergone in adaptation to an erect stance are perhaps more evident than those seen in the hand, at least within the bounds of the primates.

Man's foot is typically plantigrade. The adaptation, however, seems to be truly remarkable if one considers it to be a secondary one following that of the arboreal primates (Fig. 18.1). The foot of the latter appears to be especially adapted to climbing: it is therefore more mobile, prehensile by means of the (often incomplete) opposability of the great toe; and the axes of the heels converging downwards so that the sole appears to be turned medially. Such a configuration guarantees an excellent grip on branches and trunks, and allows the foot to exert a maximum push, thus helping the upper limb

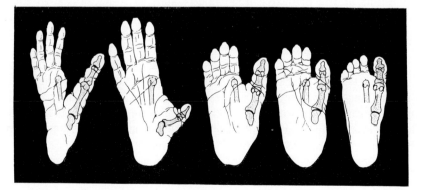

Fig. 18.1 The shape of the foot in various species of monkeys and man with special reference to the structure and position of thumbs.

efficiently during brachiation. As has been pointed out, the length of the lower limb tends to diminish as the period of time a species has brachiated increases.

The foot constitutes the only support for men's body on the ground, since his lower limb is long and columnar and the upper one shorter. Under static conditions the line of gravity of the human body must, in fact, fall within the supporting polygon formed by the feet alone. Under dynamic conditions the feet must also bear all the inertial stresses acting on the mass of the body. The human foot is, therefore, compact in structure, but at the same time elastic. The great toe is not opposable; the phalanges have little or no mobility. At the same time however, the arched sole is extremely well designed to discharge to the earth the stresses transmitted from the tibia, somewhat like the string of a long bow. The axes of the heel in humans converge in an upward direction, so that the sole, in contrast to that of the anthropoids, has a tendency to turn outwards. This provides a better lateral equilibrium on the ground, but at the same time reduces the climbing ability.

The typical structure of the human foot is ancient, but some characters, like the slight mobility of the great toe and the convergence of the axes, have been acquired only recently. In fact Neanderthal man seems to be closer to the arboreal apes than to modern man (Boule and Vallois).

3. Adaptation of the Upper Limb to Grasping with Force and Precision

The arboreal life characteristic of even the most ancient primates determined the development of particular differentiations of their upper limbs.

The upper limb lost the typical features induced by the terrestrial locomotion of the primitive quadrupeds and, as it became adapted for grasping, major structural modifications occurred. There followed the appearance of the characteristic primate "hand", which is always an organ for grasping, even when it assumes the function of support (Fig. 18.2).

The characteristics of the hand enable it to perform numerous tasks. The hand has taken over some of the functions of the buccal apparatus, such as seizing and holding food, as well as the olfactory function of recognizing objects, in this case by means of touch. All the primates use their hands in a more or less complex way; all have complete control over them. Sometimes, however, individual control of the fingers is lacking, or takes on certain specializations in different groups. In some lemurs for example, or in the American monkey *Alouatta*, objects are grasped between the second and third fingers, while in the Old World monkeys this is done much more simply, with the aid of the thumb.

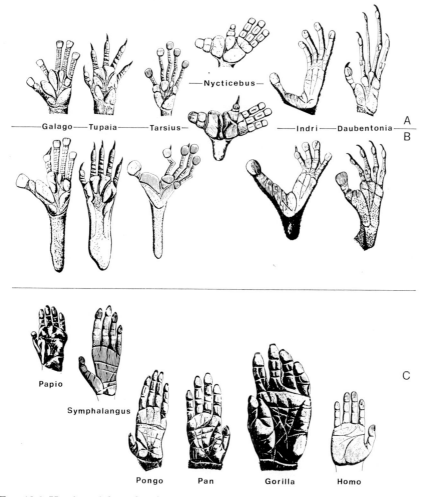

FIG. 18.2 Hands and feet of various Primates: (A) Hands of some Prosimii, (B) Feet of same, and (C) Hands of Cercopithecoidea and Hominoidea (redrawn from J. Biegert, 1964).

There are some further species (*Ateles, Colobus,* and *Symphalangus*) in which the thumb is almost non-existent.

Originally, the primate hand was functionally adapted to the animal's arboreal behaviour; the hand was used by the animal for climbing among branches. For this purpose a particularly efficient grasp is necessary, and can be made by closing the second, third, fourth, and fifth fingers around the object so as to form a hook capable of sustaining the animal's entire weight, Very often the thumb works with the other fingers, improving the grip by being more or less completely in opposition to them.

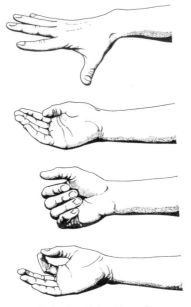

FIG. 18.3 Possible movements of the hand in Man. From top to bottom: divergence, convergence, prehensility or grasping, and opposability (after Napier, 1962).

This type of grasp is called the "power-grip", in that the muscular action is always over-abundant; it does not require fine neuro-muscular co-ordination. It can sometimes be produced simply by traction of the arm.

In contrast, when a small object is seized, the maximum force is not required and the "precision grip" is used. In this case the conscious intervention of the nervous system is indispensable in controlling the degree of muscular action, and the sense of touch and visual guidance are of the greatest importance. These types of grips are common to all the primates, but tree dwellers predominantly use the power grip, and the land dwellers with erect posture (Hominids) more often use the precision grip.

As already noted, the precision grip can be substituted in part for the sense of smell in the recognition of food and other objects. In fact, a macrosmotic quadruped presented with an unknown object will generally smell it, while a primate will touch it, turn it about and look at it. This suggests a high degree of visual and tactile sensitivity which is lacking in quadrupeds that do not have finger cushions or binocular vision.

The principal anatomical adaptations to the power grip are a short thumb which is more or less opposable, and very elongated fingers capable of a "double closure". In "double closure" the distal phalanges are arranged parallel to the metacarpals during the closure of the fist. Under these conditions the precision grip (orangutan, chimpanzee) occurs not so much by the opposition of the thumb to the other fingers, as by its placement on the tip of the index finger. This is due to the excessive length of the fingers and the shortness of the thumb. The precision grip is carried out perfectly only when the thumb is exactly opposable to all the fingers and succeeds in forming a sort of pincers with them.

The appearance of erect posture brought about an additional specialization of the hand different from that derived from arboreal adaptations. In the human hand the power grip has lost some of its force, and generally requires the complete opposition of the thumb, though this depends on the size of the object. The precision grip has become greatly refined in man because of the great sensitivity and mobility of the hand, and also because the high level of co-ordination of movement that is unique to our species.

4. Different Stages in the Evolution of the Thumb

The differentiation of the thumb can be considered as one of the most significant adaptations to take place in primate evolution. It is important in both arboreal and terrestrial life. The functional autonomy of the thumb, which is the principal factor in the capability of the human hand, did not develop suddenly in the course of evolution. Its progressive improvement has taken place during the 70 million years covered by the evolution of the primates.

The development of the thumb, from its primitive state onwards, has been divided by Napier into four stages: (1) convergence, (2) phalangeal opposability, (3) imperfect opposability and (4) true opposability (Fig. 18.3).

Comparative anatomists and palaeontologists now generally agree that the tree shew (*Tupaia*) can serve, *cum grano salis*, as an example of the very first phase of the evolution of the primates.

The hand of *Tupaia* is fairly long and the fingers are provided with pointed claws. It is able to "converge", that is to perform a movement of flexion of the metacarpal-phalangeal joints which brings the tips of the fingers closer to the centre of the palm, to form a kind of cup. The opposite movement is divergence, which opens the fingers like a fan and provides a broad carrying surface. A tree shrew is incapable of truly prehensile behaviour and it can hold objects tightly only with the help of both hands. Functionally, therefore, two hands of this type are equal to one prehensile hand. The method used by the squirrel to eat provides an example of this type of behaviour. Nevertheless, the *Tupaia* is already somewhat specialized, since it is possible to recognize the characteristics of the primate thumb in the particular disposition of its palmar muscles.

True prehensility requires, as a prerequisite, that there be a movement of the thumb not just on the plane of the palm of the hand (divergence), but also at right angles to it, so that the thumb forms a bracket with the other fingers. A similar displacement of the thumb has taken place as a parallel development in very different vertebrates, such as parrots, lizards, marsupials and primates. The latter had still to pass through two if not three further stages to develop an opposable thumb (Fig. 18.4).

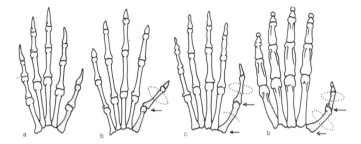

FIG. 18.4 Hands of some primates: (a) *Tupaia*, (b) *Tarsius*, (c) *Cebus*, (d) *Gorilla*. The figure shows the relationship between the evolutionary modifications of the skeletal structure and refinement of manual dexterity (indicated by movement of the thumb) (after Napier, 1962).

The first fossil documentation of prehensility is provided by the hand of *Notharctus*, a lemur of the Eocene in North America. It is particularly interesting that the great toe of this animal was clearly much more opposable than its thumb. The opposability of the thumb of *Notharctus* probably had some resemblance to that of the present-day tarsier. However, the tarsier completes the opposition, not as occurs in man by means of the carpometacarpal articulations, but through an analogous movement of the metacarpal-phalangeal articulations. In this animal these joints possess fairly free movements of adduction, abduction and rotation, which serve as a basis for prehensile activity.

This movement has been called, because of its location, "phalangeal opposability"; it is preserved in many human beings and for them plays an important role in the total mechanism of the opposition of the thumb.

The Tarsier, which is considered by many scholars to be a descendant of the common stock from which the monkeys and man are derived, displays another important specialization in its hand. Large fleshy plaques covered by thin capillary sulci are found on the volar surfaces of the tips of the fingers. Each finger also has a flat, triangular nail. The little cushions are adequately provided with sensory nerves and correspond to the pads on human fingertips.

After the phase of the tarsiers in the evolution of the thumb (phalangeal opposability), the palaeontological evidence of the successive stages is found in an epoch of 30 million years ago. In the middle Miocene of East Africa there are a number of finds which can be ascribed to the group of *Proconsul*. The hand of a *Proconsul africanus* has been almost completely preserved and has been accurately reconstructed. Although the metacarpal bone of the thumb is missing, the greater multangular bone with which it should articulate indicates that the thumb of *Proconsul* did not extend its mobility to the carpo-metacarpal articulation. However, it was not yet completely opposable. True opposability requires a saddle joint at the articular carpo-metacarpal connection.

A saddle joint permits the normal movements of flexion-extension, abduction-adduction, and also a type of combined rotation like circum-duction, so that the thumb can be moved around an axis through the centre of the articulation and be opposed to the palmar surface of the other fingers. This movement was realized only later in the course of evolution.

The "imperfect" opposability of the *Proconsul* thumb again, according to Napier, functionally resembles the level of some American monkeys (Cebidae) who can manipulate objects. The imperfect opposition of the thumb of these primates results from the fact that the greater multangular bone is turned inwards and directed towards the centre of the hand. To this is joined strong flexor tendons contained in a deep carpal canal.

5. The Acquisition of Opposability of the Thumb

The most ancient example of true opposability of the thumb in the family of Hominids is seen in the hominid fossil found in the Pleistocene layers of the Olduvai Gorge (Tanzania). Among the bones of the hand that were found was a greater multangular bone that constituted clear proof of a true opposability, and a large distal phalanx of the thumb that indicated a hand functionally similar (if not identical) to that of modern man. Another example of an opposable thumb is provided by the Swartkrans remains (Australopithecine). Among these were metacarpal bones of human type, but with a relatively short thumb. This is an indirect proof that these homi-nids were at least partially capable of manufacturing implements.

The importance of true opposability of the thumb in human evolution cannot be over-stressed. When our distant ancestors left the woods to live in more open country, and little by little began to walk on two legs, the hands began to be used. They ceased to be purely passive organs, useful only for holding firmly to branches or allowing the body to swing from trees, or for gathering food and carrying it inefficiently to the mouth. The hands became the means by which primitive man proved capable of struggling against his environment, and so becoming truly "human".

Primitive man used his hands to procure food, to hunt and to obtain

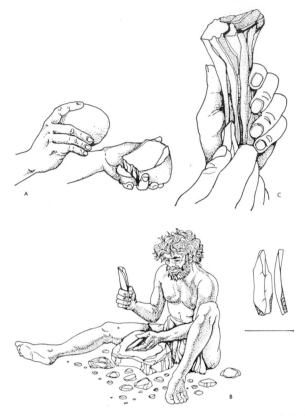

FIG. 18.5 Use of the arm in working stone (A, B) and horn (C) in the Paleolithic era. This shows the increasing dexterity of the human hand at this time.

supplies. Each of these activities requires much refinement (at the functional level) of the hand. A predator without the special teeth of the carnivores must have weapons to kill its prey and tools to cut it into pieces; a food-gatherer that does not have buccal sacs or a spacious stomach in which to store food must manufacture containers to allow for the collection and carrying of food. In succeeding epochs man used his hands to build dwellings, light fires, make clothing from the skins of animals killed for food, manufacture more highly perfected weapons and tools, and design tools with which to make other objects; finally arriving at our own age of automation in which hands are used to create machines which automatically make other objects (Fig. 18.5). At some point in the course of evolution man turned his manual and intellectual capacities from the purely utilitarian to the decorative: to paint the walls of caverns, to shape terracotta figurines and to make necklaces.

This evolutionary process can be explained only by the mechanism of natural selection which selected for the types of behaviour most advantageous to the survival of the species. With the progress of the ability to walk on two legs, weapons were perfected; that determined, as we shall see, an analogous development of the brain, especially those parts of the motor and sensory cortex concerned with the functions of the hand and the storing of sensory impressions and motor ability. The increase in cerebral capacity permitted further development of manual creative ability, from which followed a new and greater need to select for volume and differentiation of the brain. Thus, in a continuous renewal of cause and effect over hundreds of millenia, the three physical attributes which distinguish man from the other primates and make him a unique being in Nature were perfected: his erect carriage, his manual ability and, as we shall see in the next chapter, his large brain.

GENERAL REFERENCES

Biegert, J. (1964). The Evaluation of Characteristics of the Skull, Hands, and Feet for Primate Taxonomy. *In* "Classification and Human Evolution", (S. L. Washburn ed.), pp. 116–45, Methuen & Co., London.

Bishop, A. (1962). Control of the hand in lower Primates *Ann. N.Y. Acad. Sci.* **102**, 316–337.

Lisowski, P. F. (1967). Angular grow changes and comparisons in the primate talus. *Folia primatologica* 7, 81–97.

Marzke, M. W. (1971). Origin of the human hand. *Am. J. Phys. Anthrop.* **34**, 61–84.

Napier, J. R. (1960. Studies of the hands of living Primates. *Proc. Zool. Soc. Lond.* **134**, 647–657.

Napier, J. (1962). The Evolution of the Hand. *Scient. Am.* pp. 155–61.

Napier, J. and Barnicot, N. A. (eds.) (1963). "The Primates". (Symposia of the Zoological Society of London, 10). Academic Press, London and New York.

Napier, J. R. and Napier, P. H. (1967). "A Handbook of Living Primates". Academic Press, London and New York.

Oakley, K. P. (1965). "Man the Toolmaker". British Museum (Natural History), London.

Oxnard, C. E. (1969). Evolution of the human shoulder: some possible pathways. *Am. J. Phys. Anthrop.* **30**, 319–332.

Tuttle, R. H. (1970). Postural, propulsive and prehensile capabilities in the cheiridia of chimpanzees and other great apes. *In* "The Chimpanzee" (J. Bourne, ed.), Vol. 2 pp. 167–253.

19

The Evolution of the Head in Relation to Erect Posture

1. REFINEMENT OF THE SENSE OF SIGHT AND SPECIALIZATION OF THE HAND

The arboreal existence of the primates had, as its first consequences, an improvement of the sense of sight and a reduction or at least no development of the sense of smell. Rapid movement among branches and leaping from one branch to another require acute vision and an ability to judge distances accurately. Consequently, stereoscopic binocular vision became advantageous to the evolving primates and was realized as the optic axes drew nearer to the sagittal plane and thus parallel to each other. This frontal rotation of the orbit may have caused a modest increase in the size of the cranial cavity. At any rate, the change over from a "macrosmatic" encephalic structure to a "microsmatic" one, brought with it a refinement of psychic characteristics. At the same time the release of the hands from the function of support was conducive to specializations for grasping. This was achieved with the appearance of flat nails in place of the claws of more primitive primates, and with the acquisition of more or less perfect opposability of the thumb, new central nervous structures are, therefore, necessary. But if this was a stimulus for changes in the morphology of the cranium and, therefore, for an increase in its capacity, a second and more intense impetus resulted from the acquisition of erect posture.

2. Postulations Arising from the Constant Position of the Semicircular Canals

The gradual acquisition of an erect carriage caused a complex of selective reactions, particularly with regard to the direction of the earth's gravitational field. The gravitational field is an environmental component to which little attention is paid because it is practically constant, but it is certainly active in the shaping of living organisms.

The architecture of the bodies of all living beings is such as to withstand the force of gravity. A creature whose body does not adequately resist the action of gravitational forces cannot survive. The selective action of gravity is consequently very powerful.

On the way from a quadrupedal posture, in which the major axis of the body is horizontal, to an erect stance, in which the axis is vertical, substantial selective changes took place in the body in reaction to the environment (considering only its gravitational component), and modified the organism to a remarkable degree.

According to Delattre, during the time in which man evolved an erect posture, the organ of equilibrium (made up principally of the semicircular canals and the vestibule of the inner ear), must have kept its orientation, with respect to the direction of the gravitational field, constant. The three semicircular canals are placed at right angles to each other in the three planes of space, called the "vestibular planes". In practice the vestibular planes are established as follows:

1. Horizontal or fundamental vestibular plane, singled out by the external or horizontal semicircular canal.
2. Sagittal vestibular plane, identified with the sagittal plane of the cranium.
3. Frontal vestibular plane, perpendicular to the first two and passing through the centres of the circles identified by the horizontal semicircular canals. The intersection of the latter with the horizontal plane determines the so-called "vestibular axis" of Perez.

Clearly it was impossible during phylogenesis to permit the existence of any individual or species having, in conditions of normal behaviour, the fundamental vestibular plane lying in any position other than the horizontal. Such a situation obviously would be even more incompatible with the equilibrium of animals like the primates, whose motor activity has always been particularly tied to their sense of equilibrium. Consequently it must be accepted that the canals, and with them the petrous portion of the temporal bone in which they are located, must have maintained a constant position during phylogenesis while the rest of the body rotated in space.

The planes of the semicircular canals are thus a fixed element to which variations in posture occurring in the course of evolution must be referred.

3. Variations of the Cranial Bones due to Erect Posture

In quadrupedal posture the architecture of the cranium is such that the articulation of the cranium with the spine takes place along a horizontal axis. Consequently, the plane of the occipital foramen is vertical, or almost so. The head is thus united to the spine by horizontal suspension. In erect posture, the head "rests" instead on the vertebral column and the plane of the occipital foramen lies sub-horizontally (Fig. 19.1). The change from one posture to the other requires major modifications in the cranial bones since it is obviously not possible for the entire cranium to rotate through the necessary 90° without altering them.

The shift to the vertical posture in the Hominids came about in such a way that the cervical region of the column remained noticeably short, while its curvature, when present, tends towards a dorsal concavity.

These mechanisms as a whole have caused great phylogenetic variations in the architecture of the cranium, in that some parts have undergone the rotation consequent to straightening, while others, in relation to the demands of equilibrium, horizontal vision and the intake of food, have not been able to vary their position in space. This has resulted in a plastic

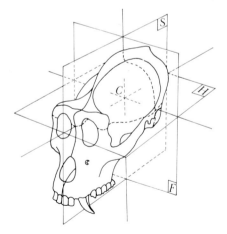

FIG. 19.1 The skull of an adult Gorilla, placed in three planes. F=frontal plane; H=horizontal plane (these two planes are placed on the vestibular axis). S=median sagittal plane or the plane of symmetry; C=central point of the head, located on the intersection of the three planes (from Delattre, 1958).

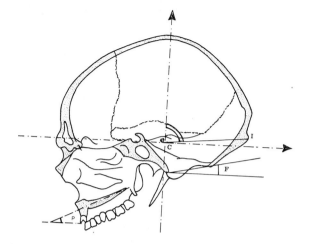

FIG. 19.2 Sagittal section of the human cranium. The dotted lines represent the vestibular bones. The angle of the foramen magnum is formed at F, that of the prolatovestibule at P, and the angle of the hiatus at CI (redrawn from Delattre, 1958).

reaction of the cranial bones, which have gradually adapted to the new conditions.

Considering the angle of the foramen or of Perez, formed by the intersection of the sagittal plane of the horizontal vestibular plane with the line passing through the basion and opisthion, one can see that it is about 90° in the carnivores and from 15° to 0° in man. The central-iniacal angle, formed by the meeting on the sagittal plane of the line joining the inion and the vestibular axis of Perez with the frontal vestibular plane (Fig. 19.1), is, on the other hand, zero or negative in the lower mammals and nearly 80–90° in modern man (Fig. 19.2).

The existing primates have intermediate angles, but their angle of Perez is rarely greater than 40°. No data exist for fossil man, since the semicircular canals have not yet been dissected in any find.

During phylogenesis, the extent of the rotation of the basal portion of the occipital bone with respect to the petrous portion of the temporal, has been in a right angle. According to Delattre the rotation of the basal portion (that is, the portion of cartilaginous origin) of the occipital bone around the vestibular axis is the evolutionary factor that has led to the actual shape of the human cranium.

This interpretation, based on the cranium of the carnivores (considered to be the prototype of primitive crania), permits interesting hypotheses and deductions to be made.

The typical carnivore cranium, orientated in the vestibular planes, is found in practice to be situated anterior to the frontal vestibular plane. The parietal-occipital suture rises obliquely towards the vertex while the petro-occipital, which is the prolongation of it, is directed downwards, partly hidden by the squamo-occipital suture. The line determined by these sutures constitutes an important point, making it possible to follow the process of transformation of the cranium during the evolution of an erect posture.

In fact, this evolutionary process developed as if, during the change to a vertical spine, the "basio-occipital" (or, according to Delattre, the palaeo-occipital) is "detached" along the above-mentioned suture line and "rotated" with a gradually increasing angle.

In the most elevated portion of the primitive parietal-occipital suture the "detachment" of the palaeo-occipital bone from the parietals takes place and the "hiatus", or virtual lacuna, which thus appears is gradually filled in by a new bone formation. This is the neo-occipital, which anatomically makes up the squamous portion of the temporal bone, and, as is known, originates in membrane. At the same time the parietals react by increasing their lateral and posterior surfaces; while the *inion* (which indicates the separation between the basio-occipital and the squamous bone,) originally contiguous with *lambda*, is gradually separated and rotated practically through 90° (see the central-iniacal angle). The rotation of lambda, determined by the extension of the parietals, appears to be about 45°.

In the lateral region adjacent to *asterion*, along a narrow portion of the parieto-palaeo-occipital suture, there is no attachment. The palaeo-occipital then tends to become larger, "dragging" part of the parietal. Thus the neo-occipital lacuna comes to be subdivided into two portions: one superior or super-asterical, the other (bilateral) inferior or sub-asterical. This lacuna will be filled by the mastoid lamina, which is well developed in the primates and is really a backward extension of the base of the petrous pyramid consequent to the rotation of the occiput. This lamina must not be confused with the *mastoid apophysis* from which it is derived.

From no part of the occipital foramen (foramen magnum) do the petro-occipital sutures undergo the process of "detachment" at the level of the *jugular processes* of the occipital bone, so that the margins of the petrous bone and the palaeo-occipital rotate and adapt together to the new conditions. It is evident from these observations that in the process of rotation from the plane of the occipital foramen, the cranium has acquired a noticeable additional surface area. This extra surface derives both from the filling in of the neo-occipital lacuna, with the formation of the squamous bone and of the mastoid lamina, and from the lateral extension of the palaeo-occipital bone. Obviously such an enlargement of the surface area corresponds to an increase in the mass and functioning of the brain.

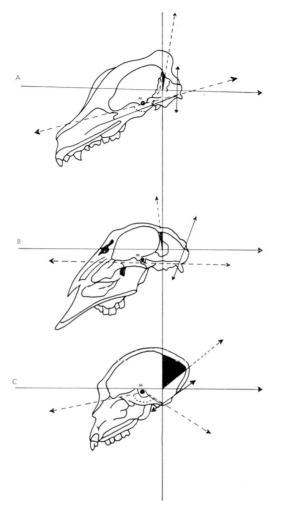

FIG. 19.3 Movement of the base of the skull of some different types of Mammals. "M" indicates the position of the sella turcica. The hiatus is indicated by the shaded area; the direction of the foramen magnum is indicated by a double arrow in full line (redrawn from Delattre, 1958).

4. REACTION OF THE CRANIAL STRUCTURES TO THE NEW CONDITIONS

The base of the cranium, which in its primitive condition appears elongated, rectilinear and downward sloping (carnivores), partially follows the occiput in its rotation. While in the occipital bone the iniacal region is lowered, in the basilar region the clivus tends progressively to be raised.

This mechanism can be explained, keeping in mind the *sphenoidal* angle (basion-central pituitary-posterior cribriform point) which expresses the relative positions of the *clivus* and the *planum*. According to Delattre's model (Fig. 19.3) (acceptable as a mechanical variation but not appropriate for showing phylogenetic relationships), in the transition from the primitive architecture of *Canis* to the more differentiated type of *Ovis*, there is, with a slight rotation of the foramen magnum, a total raising of the cranial base. The sphenoidal angle, in fact, remains about 180°.

In the primates, along with a more accentuated rotation of the occipital, a "fracture" appears between *planum* and *clivus*. Thus angles are obtained of less than 180° (about 140° in the macaque and 120° in man). The result is that the root of the nose moves closer to the anterior margin of the occipital foramen, causing a change in the position of the face with respect to the neuro-cranium.

Since it causes variations in the reciprocal relationships between the bones, the rotation of the occipital bone determines major modifications even in bones far away from it. The parietals double their surface area and extend upwards and backwards. The temporal bones turn around the vestibular axis while the squamous portion increases in size. During this movement the horizontal semi-circular canal obviously maintains a constant position. The increased surface of the bones of the cranial vault provides a larger volume for cranial activity, but since the vault is elongated and raised, the shape becomes progressively more spherical. Due to the geometrical properties of this form the increase in volume is proportionately more rapid than the increase in surface area.

It must be kept in mind that phylogenetically the spheroidal form appears very late, practically only in *H. sapiens fossilis*. Before this the purse shape (birsoides of Sergi) is universal. This shape is typical of all Neanderthal men and of all *H. erectus*, as well as the greater part of the existing and fossil higher primates. Thus a surprising delay existed between the acquisition of erect posture and the assumption by the cranium of a globular shape, while the acquisition of greater volume (1500–1600cc) preceded the latter occurrence. Neanderthal man has a capacity corresponding to the average of modern man (crania of La Chapelle-aux-Saints and Circeo); but his cranium is extremely disharmonious. The parietals and the squamous portion of the occipital bone appear to be completely developed, while the frontal bone is narrow and receding, with wide supra-orbital ridges. The skull-cap is therefore low, wide and elongated.

Probably this remarkably disharmonious architecture should, morphologically and statically, be considered in relation to the retained notable size of the splanchnocranium, with the consequent bulkiness of the masticatory and neck musculature. Only with the improvement of the conditions of life and the environment, partly due to the post-glacial climate and partly to cultural advances, did a selective process which tended towards slimming of cranial bones and head become possible. Such a process resulted in a harmonious redistribution of the mass, with reduction of the face, verticalization of the frontal bone and an increase in the size of the skull-cap (Fig. 19.4).

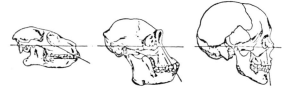

FIG. 19.4 Progressive reduction of the muzzle from Mammals to Man, as demonstrated by the change in the facial angle.

An interesting example of these changes is the variation in the angle of the chin which increases during the hominization process (Fig. 19.5). This same variation occurs during human ontogenesis. The meaning of this change is yet unknown but there is speculation that it is related to the general re-organization of the face following the process of reduction of the splanchnocranium. This change occurred due to changed alimentary habit and the shifting of the foramen magnum from a posterior to a medial position.

5. ENSUING EVOLUTION OF PSYCHIC CHARACTERS

The rotation of the foramen magnum around the vestibular axis led to an increased cranial capacity and this in turn produced greater possibilities for the development of the brain. The simplest aspect of this development was the increase of the cerebral mass.

Capacity in itself, however, bears only a crude relationship to an increase in psychic activity. More important, as we shall see in the next chapter, are the distribution of the encephalon mass, and the possibilities of development of the cerebral cortex. These are realized when the surface of the lateral hemispheres is increased. The larger volume at the disposition of the brain permits more complex convolutions, a further increase in the cortical surface, and major variations in the proportions of the various parts. For example, the prefrontal bone occupies 8·3% of the cerebral surface in the lemurs, 12% in the Cercopithecines, 16·9% in the chimpanzee and a good 29% in man. To summarize, it can be stated that the process of acquiring erect posture led directly to the development of the neopallium, the most phylogenetically recent part of the brain.

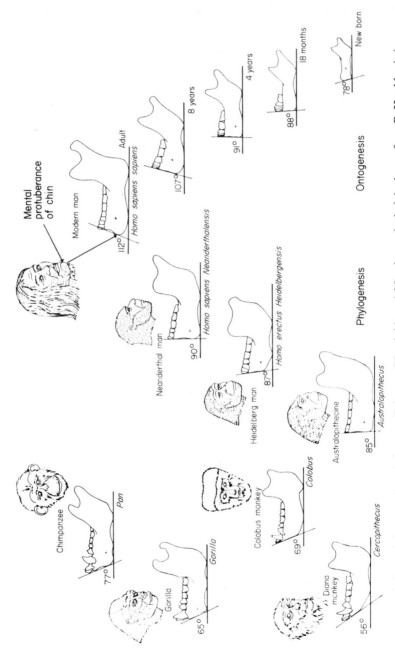

Fig. 19.5 Variations of the mental angle in some Primates (Hominids and Man in particular) (redrawn from P. Hershkovitz).

Unfortunately, it is impossible to determine what the morphology of the brains of fossil hominids was like, even though some information can be derived from endocranial morphology. It is, however, a fact that, when the capacity increased and the volume was differently distributed, the psychic manifestations of the hominids, evidenced by the objects they manufactured, changed considerably.

6. Variations in the Musculature of the Head

The acquisition of an erect posture brought about profound modifications in the musculature of the head. These occurred due to the new conditions of mechanical equilibrium, the prehensile and masticatory activities of the buccal apparatus, and the heightened importance of mimicry. The rotation of the palaeo-occipital bone, previously described, produced a different static condition of the head. In fact, in man the head articulates *above* the vertebral column in the fashion of an architrave, while in quadrupeds it is articulated at the end of a "vertical beam". The physiological (and gravitational) horizontal position can thus be maintained with less expenditure of muscular energy in man than in the quadrupeds. This is true insofar as the occipital portion acts as a "counterweight" to the facial portion with respect to the neurocranium, as happens in modern man and, to a decreasing extent, in the Neanderthalers; the Pithecanthropi and in the apes which have a sloping posture (clinogrades). In the quadrupeds, however, *all* the weight of the head must be counterbalanced exclusively by the action of the muscles (Fig. 19.6).

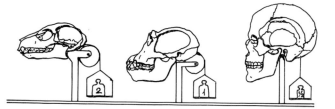

Fig. 19.6 Statics of the skull: position of the foramen magnum and the action of the counter-balance of the occiput. From left to right: quadruped (*Canis*), Pongidae (*Gorilla*), and modern man.

These variations can easily be recognized in the skeleton; the reduction of the nuchal line and of the external occipital protuberance as a result of the reduction of the muscles in the nuchal region, is one example. Still, parallel to this reduction there is a greater development of the mastoid apophysis, which is insignificant in the non-human primates. This is due to mechanical demands arising from the lateral equilibrium of the head, so that there is a remarkable development of the sternocleidomastoid muscle, which in the quadrupeds has little or no function.

Likewise, a reduction is observed in the temporal muscle, which in quadrupeds and in some of the higher primates (e.g., the gorilla) is inserted in the sagittal parietal crest. The crest is therefore absent in man. In modern man this muscle originates laterally at the level of the temporal lines and only slightly invades the region of the lateral margins of the parietal and frontal bones. The reduction is largely due to the reduced dimensions of the dental arch and the refinement of the mandible. A robust masticatory apparatus is not a positive character for modern man's survival and, therefore, does not have positive selective value. Along with the temporal muscle, the masseter is reduced, and consequently so are the zygomatic arch and the ascending rami of the mandible.

Linked to the increase in psychic activity is the development and perfection of the muscles of facial expression, while those of the neurocranium are reduced.

GENERAL REFERENCES

Editorial (1958). "Les Processus de l'Hominisation" (Colloques du C.N.R.S.), Editions du Centre National de la Recherche Scientifique, Paris.

Coon, C. S. (1962). "The Origin of Races", Alfred A. Knopf, New York.

Delattre, M. A. (1951). "Du Crane animal au Crane humain", Masson et cie, Paris.

Delattre, M. A. and Fenart, R. M. A. (1960). "L'hominisation du crane", Editions du Centre National de la Recherche Scientifique, Paris.

20

Cranial Capacity and Its Growth from Primates to Man: the Differentiation and Evolution of the Brain

1. HUMAN CRANIAL CAPACITY AND ITS VARIATIONS

The cranial capacity of human beings averages about 1400cc, with a maximum of 2000cc. Women have a cranial capacity which is slightly less than that found in men. The difference is, on the average, about 150cc and the female capacity corresponds very roughly to a value between 86% and 95% of the capacity of the male skull. The sexual difference exists from the time of birth. Cranial capacity increases in both sexes up until about twenty years of age. To give an idea of the relative value of the development of the cranial capacity it is noted that at birth the value is slightly more than one-quarter (370–380cc) of the adult value, while the body weight of the new-born is about one-twentieth of his adult weight.

The lowest values for cranial capacity among human populations have been found among the Tasmanians (now extinct), the Australian aborigines (1250–1300cc), the Andaman Islanders, the Vedda, and the Bushmen in Africa. The highest values have been found among the Mongolians, among whom cranial capacities greater than 1500cc have been found, however, no definite relationship has been established between cranial capacity and intellectual development. Some scientists, writers and artists like Kant, Byron and Volta have had remarkably large cranial capacities, but they must be compared with Foscolo and Raphael who had cranial capacities that were far below average.

2. Cranial Capacities of the Fossil Hominoids and the Anthropoids

Among the fossil remains that can be attributed with certainty to our species cranial capacities are found similar to, and perhaps even greater than, those seen in living groups of man. Thus, in Cro-Magnon man capacities around 1600cc have been found: the cranial capacity of Chancelade man surpassed 1500cc; Neanderthal man, living in the middle Palaeolithic, had an average a little below 1500cc. Among the fossil Phaneranthropi, that of Fontechevade had a capacity of about 1470cc.

Pithecanthropus and *Sinanthropus* had cranial capacities decidedly inferior to the above-mentioned averages, although not lower than those of some present-day populations. According to Lahovary, an average of 900–1000cc can be assumed for *Pithecanthropus* and one of 1000–1100 for *Sinanthropus*.

The Australopithecines had a decidedly low cranial capacity, falling between 435 and 650cc. Thus, it was below the value of 850cc which Vallois indicates as the minimum possible value for human beings (Fig. 20.1).

The anthropoid apes have still lower cranial capacities. The cranial capacity of the male gorilla averages about 535cc (with a maximum of 752cc); while that of the male chimpanzee is 381cc, with maximum of 454cc. The cranial capacity of the orangutan is about 400cc and that of the gibbon about 100cc (Fig. 20.2 and Table 20.1).

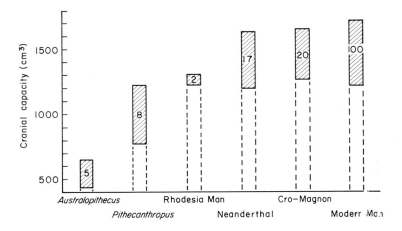

Fig. 20.1 Increase in brain (encephalic) capacity from the most ancient hominids to modern man. (The numbers indicate the number of specimens for which the capacity was calculated.)

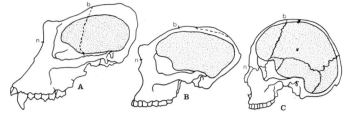

FIG. 20.2 Comparison from a lateral view of the cranial capacity of *Gorilla* (A), *Pithecanthropus* (B), and *Homo* (C).

3. RELATIVE ENCEPHALIC WEIGHT

However, "cranial capacity" is not a relevant concept when it is considered by itself as merely a cavity. One must also take into account the cerebral substance contained in this cavity and its relationship to the body mass. It is therefore necessary to seek an expression which accurately evaluates the assumed development of the mass of the brain in relation to total body mass. Such an expression is the "relative encephalic weight". One simple method consists of expressing the weight of the brain as a percentage of the weight of the body. In this way the number taken as the "index of cephalization" can be obtained.

The application of this calculation to the zoological scale has furnished the following gradation of values: Man 2·17; Mouse 2·04; Squirrel 1·53; Gibbon 1·37. Cat 0·94; Macaque 0·59; Chimpanzee 0·54; Dog 0·27; Lion 0·18; Elephant 0·18; Horse 0·14; Whale 0·01.

It is obvious that the values of this index of cephalization do not correspond to what is commonly accepted as the psychic development of the different species of animals, but favours in an almost indiscriminate way (with the exception of man) animals of smaller body size.

To devise a more satisfactory index the problem can be attacked *a posteriori*, by trying to determine what the relationship is between encephalic weight and somatic weight. With this new coefficient of cephalization the preceding series can be reorganized as follows: Man 2·82; Elephant 1·24; Gibbon 0·76; Chimpanzee 0·74; Horse 0·45; Macaque 0·36; Cat 0·32; Lion 0·30; Dog 0·29; Whale 0·26; Squirrel 0·21; Mouse 0·07. As can be seen this scale of values is to some degree an improvement on the preceding one.

Studies on the dimensions of the brain in different groups of mammals have demonstrated that these dimensions do not increase in direct proportion to bodily dimensions, but according to a power of about 0·66. In this way, if the dimensions of the body are doubled, the dimensions of the brain increase by $2^{0.66} = 1.6$ times. It is interesting to observe that this relationship is almost the same as that which links the increase in body surface with body dimensions, so that it seems that there is a constant relationship between the dimensions of the encephalon and body surface.

TABLE 20.1. Ranges of variations and averages of the cranial capacity (in cm³) of adult hominoids (from Schultz, 1965).

Genus	Sex	Specimens	Range of Variations	Average
Hylobates	♀	86	82–116	101
	♂	95	89–125	104
Pongo	♀	52	276–425	338
	♂	57	334–502	416
Pan	♀	57	282–415	350
	♂	56	292–454	381
Gorilla	♀	43	350–523	443
	♂	72	412–752	535
Homo	♀	50	1129–1510	1330
	♂	92	1246–1685	1446

Body dimensions in relation to encephalic ones have recently been re-examined by Jevison for many primates. The relatively strong position of the anthropoid apes with respect to man (Table 20.2) can be explained by their large body dimensions compared with those of the other monkeys. Nevertheless these attempts to quantitate the evolution of the dimensions of the brain do not take into account the differential development of the various parts of the brain in the animal hierarchy, especially in the primate species, nor is it concerned with the variation in the number of neurons and their complex interrelations. These aspects are probably the most important and, to date, the most neglected in studies of primate brain evolution.

TABLE 20.2. Index of cephalization according to Jevison (1955).

Index of cephalization: $K = \dfrac{\text{encephalic wt}}{\text{body wt } 0.66}$	
Mammals (primates excluded) ($n = 108$)	0.06
Anthropoid Apes ($n = 35$)	0.29
Platyrrhines and Catarrhines ($n = 50$)	0.41
Modern Man ($n = 50$)	0.92

n = number of individuals studied

4. GENERALIZATIONS ON THE EVOLUTION OF THE BRAIN IN THE ANIMAL SERIES

In order to study the evolution of the human brain it is necessary to take into consideration the series of animal encephala which anticipate, in some aspects, those of human beings. Since we have no fossil brains, we must use the endocranial cast which, in many species of animals, is the negative image of the shape of the brain.

An endocranial cast gives us information not only regarding the total volume of the organ and its parts, but also on the topographical relationships between the cerebral zones. This in turn, makes it possible to determine the presumed extension of the peripheral innervation of each zone. However, other, and in some respects more valuable, information can be obtained from the comparative study of the morphology of the brains of the various forms of vertebrates. The Teleosts have what can be considered a primitive form of vertebrate brain. Ignoring differences due to ecological and ethological peculiarities, it is possible to observe in a typical teleost (for example, the pike) first, a prosencephalic region which is essentially olfactory and characterized by large olfactory bulbs. This is followed by the cerebral hemispheres, which are very large but made up exclusively of the basal nuclei covered by a thin ependymal sheet which takes the place of the pallium. (The pallium is the telencephalic structure which will develop later in the classes of the tetrapods.) The diencephalon can barely be seen dorsally; but even at this level, it shows a certain degree of complexity in the organization of both the dorsal (sensory) and the ventral (motor) areas. Special structures are seen in the epithalamus (parietal formations) and in the hypothalamus (vascular sac) (Fig. 20.3).

The mesencephalon is the optic region; in its upper portion it consists of two conspicuous "optic lobes". The optic roof comes to constitute the vault of a wide mesencephalic ventricle, occupied in part by an evagination of the cerebellum, called the "valve". Then comes the rhombencephalon, represented dorsally by the cerebellum and ventrally by the medulla oblongata. The latter is clearly visible in a vestibular-lateral area, a portion which is joined to the eighth nerve and that of the lateral line which receives sounds and mechanical (static and dynamic) stimuli. The ninth, gustatory nerve is also connected, and part of the tenth, concerned with non-specific visceral sensitivity.

The medial area of the rhombencephalon, which continues in the ventral part of the mesencephalon, is the site of both visceral and somatic motor centres in connection with the real origin of the cranial nerves and the long descending pathways of the spinal cord.

The cerebellum, the organ which has the function in all vertebrates of

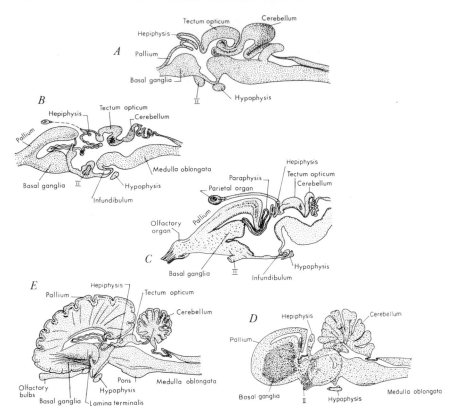

FIG. 20.3 Schematic representation of the cerebrum of five different kinds of vertebrates, all drawn to scale. (A=Teleostei, B=Amphibia, C=Reptilia, D=Aves, E=Mammalia.

co-ordinating movements, is of particular significance because its great development (which was already mentioned in connection with the valve which protrudes into the mesencephalic ventricle), is due to the imposing collection of proprioceptive fibres coming from the lateral line and the vestibule.

In the Amphibians the hemispheres are narrow, prolonged rostrally in the olfactory bulbs, and project backwards to cover part of the diencephalon. Their vaults are thickened and constitute the pallear area of the telencephalon, which is mainly concerned with olfactory sensation. Ascending fibres from the diencephalic and rhombomesencephalic segments are already present.

In all the telencephalon the grey matter is rather deep around the ventricles: only in the dorso-medial pallium (the rudimentary hippocampus and archipallium) do the cells diffuse over the full depth. Even then they do not achieve the regular stratification of a true cortex, nor foreshadow its evolution.

The mesencephalon (optic lobes) make up the most conspicuous formation of the encephalon.

The cerebellum, on the other hand, is reduced to a fairly thin sheet. In fact, the accentuated development of the cerebellum in fish is justified, as we have seen, by the efficiency of the organs of the lateral line; but in the tetrapods, in which that system has disappeared, the cerebellum undergoes a lesser development. Still at this stage, in the amphibians, and later in the forms developing an amnion, it comes to assume major importance for equilibrium and motor co-ordination. In the tetrapods proprioceptive stimuli coming from the muscles and tendons of the trunk and the limbs reach the cerebellum, which is oriented to these stimuli in its evolution. In the more evolved mammals, and hence in the primates and in man, the development of the cerebellum is thus correlated with the function of directing and co-ordinating voluntary movements. These have their neuro-anatomical basis in that motor system which is established and strengthened to an unprecedented degree in the mammals: i.e. in the pyramidal tracts, which from the neocortex reaches the spinal cord and the effector centres.

In the Reptiles the two hemispheres, more or less pedunculate, present a pallium which is thickened as much or more than that of the amphibians, and is more extensive both anterior-posteriorly and laterally. A fair cyto-architectural arrangement shows that the grey matter is already spread out in remarkably extensive areas, but does not yet reach the complexity of a true cortex.

Ventrally and medially we find the septum flanked by a wide striated region which foreshadows that of the birds and mammals. Three regions can be distinguished: a ventral palaeostratum, above which are a neostriatum and an archistriatum.

In the pallium, which lies immediately above the septum, we find a layer of granular cells regularly arranged to form the fascia dentata; laterally to this we find a layer of pyramid cells, called the hippocampus. The fascia dentata and hippocampus constitute an ulterior evolution of the archipallium already described in the amphibians and corresponding to the ammonice formation in mammals.

The region of the lateral pallium on the other hand represents a palaeopallium, which is homologous to the dorso-lateral region of the telencephalon of the amphibians. Its grey matter has a suggestion of stratification into a palaeo-cortex.

Between the archicortex (dorso-medial pallium) and the palaeocortex is inserted an area which extends even farther into the frontal region of the hemispheres. It represents the first beginnings of a structure which is greatly developed in the mammals, and which forms the neopallium and the neocortex; in other words it is the rudimentary neocortex. The importance of the appearance of this new pallial area lies in the fact that while the palaeopallium and the archipallium remain pre-eminently olfactory centres, the neocortex is of much greater significance. It is concerned with voluntary movement through the origin of the pyramidal tracts, with association and general somataesthesic perception, and with special sensations which arrive from the thalamic projection.

In the Mammals, then, the hemispheres become steadily larger, because of the spreading out of the neocortex. This also requires an increase in the underlying white matter, that is, the fibres which arrive in the neo-cortex from the diencephalon form associations between the various areas and then make up the voluntary motor pathways.

As has been indicated above it is essential to relate the development of the telencephalon with the formation of the cerebellar hemispheres and the neocerebellum. The cortex of the latter is connected with the cerebral neocortex by means of the corticopontocerebellar path, with the interposition of the centre of the pons Varolii.

In the Mammals the two hemispheres are functionally joined by three comparable systems: one for the pallium, the anterior palaeopallial commissure; one for the hippocampus, archipallial, and the corpus callosum for the neocortical areas. The progressive evolution of the neocortical structure requires a hyperdevelopment of the corpus callosum, which in turn involves a readjustment of many telencephalic structures (reduction of the dorsal hippocampus to sterile striae, formation of the trigonum with the psalterium and the fornix, etc.). The interpretation of the evolution of the archipallial and neocortical formations from the reptiles to man was achieved through the study of mammals lacking massive corpora callosa (monotremes, marsupials), and of those primitive creatures in which the neocortex and the commissure of the corpus callosum gradually and progressively acquired greater importance.

Underlying the pallial formations are found the basal nuclei, the structures of the septum, and the corpus striatum.

The septal formations which are highly developed in primitive mammals, are reduced in primates and in man to the two thin sheets of the septum lucidum.

The striated formations represented by the nucleus are always conspicuous and are represented by the caudatus and the lenticular nucleus. They are situated in positions ventral to the pallium, which extends all around them, and lateral to the thalamic masses.

In the pallium one can distinguish:

(a) an archipallium, corresponding to the archipallium of the reptiles and of the amphibians (precursor of the hippocampus). In the mammals it is called the ammonice formation. The archicortex, that is, the cortex of the ammonice has a cytoarchitecture with a fascia dentata and hippocampus of evident reptilian derivation;

(b) a palaeopallium, corresponding to the palaeopallium of the reptiles and amphibians;

(c) a neopallium, which is inserted between the archi- and palaeopallium becomes more fully developed, and finally takes up the greater part of the cerebral mantle—the neocortex.

The archi- and palaeopallia, except in the insectivores (and in more primitive brains) are less conspicuous structures than the neopallium, and maintain connections with the olfactory organ. The telencephalon must, as has already been mentioned, be considered as the primitive olfactory region of the encephalon. It is therefore logical that in the mammals its most ancient portion maintains its relationship with that sense. The more developed the telencephalon is the more refined is the sense of smell. The archi-/palaeopallium make up the rhinencephalon, and, stimulated by the neocortical development of conspicuous masses of nervous tissue in the ventral region of the telencephalon, they elongate rostrally as the olfactory bulbs. They are separated by means of a deep sulcus, the rhinal fissure, from the neopallium which is dorsal and lateral to them.

In the microsmatic mammals (that is the mammals with much-reduced olfactory organs, such as human beings and monkeys), the rhinencephalon is greatly reduced in size.

The Neopallium

In some primitive mammals, for example the insectivores and the "edentata" which have a low level of psychic development, the development of the neopallium is less than that of the rhinencephalon.

Normally it contributes the greater part of the tissue of the hemispheres. It has a neocortex of very precise cytoarchitecture and a more or less furrowed and folded surface, proportional to bodily mass and the development of mental processes and activities.

The neocortex can be divided into a number of regions, by employing a criterion that is partly topographical, partly cytoarchitectural, and partly functional (neurophysiological).

As we have already indicated, three types of neocortex can be distinguished functionally: sensory, motor and associative.

The afferent pathways, which are essentially of thalamic origin, reach the sensory cortex carrying visceral and somatic sensory stimuli of both general

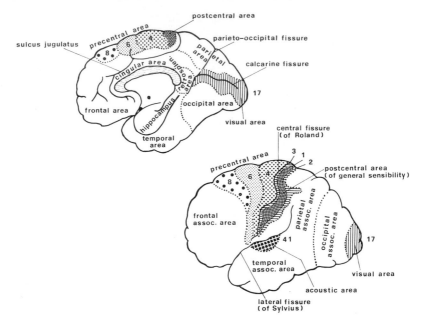

FIG. 20.4 Human brain with localization of sensory and motor specific area.

and specialized sensations. The motor impulses leave the motor areas by means of corticofugal fibres. In the association areas the various senses are integrated and memorized and here, in man, arise and are established self-knowledge, reasoning and consciousness.

Sensory areas concerned with general sensation, visual areas, auditory areas, and motor areas are found in all the mammals up to man. Little by little the psychic functions have become refined and the associative areas steadily larger.

The following regions have been distinguished in man: an occiptal visual area, or area 17 of Brodman; a temporal auditory area, areas 41, 42, 22; a (probably) gustatory area located in the region of the insula; a post-central somataesthetic area (because it is situated on the lateral surface of the hemispheres, behind the central sulcus or fissure of Rolando) or areas 1, 3, 5 in Brodman's classification; and pre-central motor areas, areas 4, 6 and 8 (Fig. 20.4).

5. CEREBRAL DIFFERENTIATION AND LOCALIZATION OF THE CORTICAL CENTRES

The recent development of neurosurgical techniques has made it possible to extend our knowledge of the localization of the functions of the human brain. According to these findings, for example, only in man does the cerebral cortex, or more precisely Broca's area, have a mechanism for the control of the utterance of sounds; in the other mammals sounds are controlled by a completely different part of the brain.

In the transition from the anthropoid apes to man there is an enormous increase in the extent of the temporo-parietal cortex.

The search for homologues in the cortical centres among the various species of primates and man is one of the most fascinating fields of neurophysiology, but to date few data are available in this field. It is certain, however, that in man there exist exclusively human centres.

How has it been possible for centres capable of discharging special and unusual functions to arise in the human brain, if they formerly did not exist ? The answer probably lies in the plasticity of this organ. Penfield and Roberts (1959) state that "human cerebral hemispheres are never equal, either as regards form or the disposition of the convolutions and sulci".

In the majority of human beings, for example, the cortical area concerned with speech lies asymmetrically in the left cerebral hemisphere, but may also be found in the right. It is probably due to this plasticity of the brain that there is the continuing possibility of the evolutionary improvement of our species.

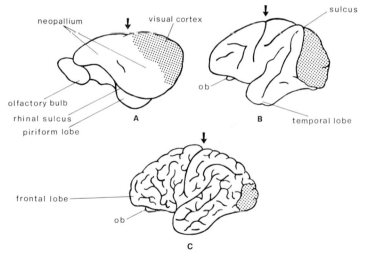

FIG. 20.5 Cerebral hemispheres of treeshrew (*Tupaia*) (A), *Macaca* (B), and man (C), all drawn to the same size. Note the variation in extent of folding of the cortex and the relative size of the olfactory bulbs (ob) and visual cortex (shaded). The temporal and frontal lobes of the neopallium hide the periform lobe of the archipallium in the higher primates, which is visible here only in the brain of *Tupaia*. The arrow points in each drawing to the central sulcus, which separates the frontal and parietal lobes. The rhinal sulcus separates the archipallium and neopallium and the lunate, or "simian" sulcus demarcates the visual cortex (after Campbell, 1969).

The Tupaioidea seem to represent the connecting link between the insectivores and the prosimians. The brain of *Tupaia* is more differentiated than that of the insectivores (Fig. 20.5) but because of differences in the characteristics of body and of extremities, does not show any of the traits specific to the brains of the primates. The *Tupaia* brain is very similar to that of some arboreal rodents and may be considered as a transitory form.

The original form of the brain of the lemurs is still not known. Recent species of this group show a highly variable level of evolution, with very primitive forms (*Microcebus*) and more specialized species (*Daubentonia, Propliopithecus*) which, however, do not reach the level of the monkeys.

The most primitive form of brain among the monkeys is found in the *Callithrix*. It differs from that of the lemurs in that the rhinencephalon is substantially reduced; the telencephalon forms an occipital angle; and there is a more definite fissure of Sylvius. The neopallia of all the higher monkeys (Fig. 20.5) are more grooved; the frontal, parietal and temporal lobes are more clearly evident. With the progressive extension of the cerebral mantle there is a gradual diminution of the sensory centres, while the association areas which form between them increase and become more pronounced.

The evolution of the encephalon of the anthropoid apes and of man is well documented by many fossil finds and by their endocranial casts. Without corresponding examples in other phyletic lines, one might say that in the hominids the outward differentiation of the encephalon and the increase in its volume happened almost explosively in a comparatively brief lapse of time (half a million years), and without any notable increase in bodily dimensions.

The morphological changes which took place can be recapitulated in the following phrases:

1. Expansion of the anterior part of the brain and especially of the neopallium.

2. Increase in the number of fissures in the neopallial cortex, with the appearance in particular of the true fissure of Sylvius.

3. Expansion of the occipital lobes, which, increasing in the posterior portion, are distinguished from the adjacent parietal areas by the formation of a post-calcarine sulcus. This lobe interprets and correlates visual data.

4. Elaboration of a precentral cortex. The frontal lobe is responsible for the muscular control of the vocal apparatus; the prefrontal lobe is important as an association area.

5. Elaboration of the temporal sulcus, especially in the more evolved

primates. This is associated with the perfecting of the ability to discriminate between sounds required for vocal communication.

6. Elaboration of the cerebellum and the installation of its connections with the motor area of the cerebral cortex.

7. Reduction of the neural effector mechanisms as, for example, the rhinencephalon.

The general effect of the increased convolutions of the cerebral cortex in the primates has been to subordinate the control of the lower centres of the brain. This is especially evident in the centres which control movements. If the entire cerebral cortex of a cat or dog is removed, the animal is still able to stand on its feet and walk almost as properly as a normal animal, thanks to the activity of the extra-pyramidal involuntary motor centres.

In the primates, on the contrary, the precentral cortex has taken over the greater part of the control of the skeletal muscles, and almost total paralysis follows the resection of the cerebral hemispheres of a monkey.

6. An Attempt at a Functional Interpretation of the Evolution of the Brain of the Primates

Life in the trees implies a complete consciousness of the heterogeneity of the environment. This requires an increase in the elaboration of external and proprioceptive perception at the level of the sense organs. In the evolutionary hierarchy of the primates, therefore, there is present a progressive and harmonious increase of those parts of the cerebral cortex which are connected with the sensory organs. Moreover, because perfect and co-ordinated control of movements and equilibrium is essential in such a precarious habitat, the cortical areas which control movements and the cerebellum are highly developed.

The result of all these exigencies is an increase in the cerebral mass. This factor is already evident in the most primitive and ancient forms of primates such as *Adapis*, when compared with the insectivores. The enlargement tends to progress throughout the entire order, culminating in man.

By means of neuro-anatomical investigations, electrical stimulation, and experimentation by means of resection of parts of the living brain, it is possible in many cases to identify the functions of these different parts.

In Fig. 20.5 the most important functional areas of the cerebral cortices of *Tupaia*, macaque and man are represented. The evolution of these different areas (olfactory, visual and auditory) in the primates can be related to the differentiation of the sense of sight and to the evolution of the sense of hearing, which in the primates is associated with equilibrium.

We shall now begin to trace how these sense organs were transformed and to consider the consequences of the transformation.

Sight

The importance of sight in the evolution of the primates is obvious: it is in direct relation to the forest environment in which the primates evolved and to the arboreal life which distinguished them from the other mammals. Capable of receiving radiant energy of between 380 and 760 nm, the eyes gather information across space regarding the dimensions, colour, movement, form, spatial relationships, and distance of objects.

There are three parts of the eye which must be considered separately in the evolution of sight (1) the constitution of the retina (2) the neural correlation of sight and (3) the transformation of the orbital region of the cranium. At this time, the third aspect is of little interest. The first two are more important for the interpretation of the evolution of the encephalon.

There are species, including some primates (prosimians), which have particularly well developed nocturnal vision. This implies a high density of photo-receptors for short length waves (red) in the retina.

The amount of information received by the brain through the eyes is, however, related to the dimensions of the optic nerve or, rather, to the number of its fibres. According to van Bonin (1963) the dog and cat have about 150,000 fibres in each nerve; man has 1,200,000.

The differences between man's eye and that of the monkeys in general are minimal. Both have a retina rich in "cones", specialized for interpretation, rather than for simple perception, of an object.

This characteristic, which permits us to read, is due to the arboreal life which our monkey-like ancestors led for millions of years.

However, the event that is of the greatest interest in understanding the evolution of the part of the encephalon connected with vision is the evolution of stereoscopic vision. In the primates this required a change in the position of the eyes, from lateral to frontal, and also changes in the structure and organization of the optic nerve and thus of the brain.

In man, in fact, about half of the fibres of the optic nerve which come from each eye pass to the same side of the encephalon (Fig. 20.6). The information gathered by one eye is, in this way, distributed equally to both parts of the brain.

In fact, the evolution of the vision of the primates implies a vast expansion of that part of the encephalon responsible for the elaboration of visual perception (the visual cortex of the cerebral hemisphere).

As is evident in Fig. 20.5, the cerebral cortex concerned with vision is already very well developed in *Tupaia*, and this area has remained constant

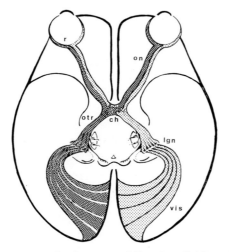

FIG. 20.6 Diagram of human visual system: "r" retina; "on" optic nerve; "ch", optic chiasma; "otr", optic tracts; "lgn", lateral geniculate body, and "vis", tract to the visual cortex. Fibres originating in the inner or nasal halves of the retinae cross at the chiasma so the visual cortex of each side receives signals from the same half of the field of view; that is, the opposite half. The signals, however, represent two slightly different images, and on the basis of these differences depth and distance are perceived (after Polyal, 1957).

in all the primates. (It must be kept in mind that in the figure the encephala of the macaque and of man must be enlarged to be in proportion.) If we compare the visual area of the cerebral cortex of *Tupaia* with that of any insectivore, we see an enormous difference. Nevertheless in *Tupaia* the sense of sight has not yet completely replaced the sense of smell, as can be seen from the large dimensions of the olfactory bulb. In the more advanced primates this substitution has already taken place, even if the visual area seems smaller in relation to the general enlargement of the brain.

Smell

That portion of sensitive integument found on the external surface of the nose (*rhinarium*) is already a well developed organ in the prosimians. The moisture which covers it is certainly very important in increasing its sensitivity and establishing the direction of the stimuli.

In primates, in whom the air needed for respiration passes through the nose and over the mucous membrane containing the olfactory receptors, the sense of smell is much reduced, even if it is somewhat important to present-day humanity. Olfactory function is, in fact, in strict relation to feeding and sexual stimulation, In modern man odours have great importance as a stimulant to appetite, but an individual whose sense of smell has been destroyed can survive without difficulty if he eats in a good restaurant or at home.

In almost all the primates the sexual stimulus is largely visual, and odour plays a secondary role, even though modern human females tend to use more and more perfume as a sexual stimulant.

The relative dimensions of the olfactory bulbs of the brain, where the olfactory nerves arrive, are shown in Fig. 20.5.

The analysis of olfactory impulses begins in the olfactory bulb. Nerve fibres go from it to the inferior lateral part of the cerebral hemisphere (piriform lobe), where presumably further association takes place.

Hearing

The ear and its associated organs, form a complex receiving mechanism, sensitive to sound-waves, gravity and movement.

Like light, sound waves move in straight lines and therefore can be used to indicate direction. For this reason the ears are separated (as are the eyes but not the olfactory organs) and can function as a stereophonic organ. They can supply information on the direction, the frequency and the amplitude of the sound waves.

The ear of a primate is divided into: the external ear, middle ear, internal ear, utricle and semicircular canals (Fig. 20.7).

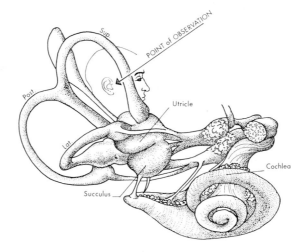

Fig. 20.7 The mechanoreceptors of the inner ear are a system of interconnected liquid-filled tubes. They are shown here with their nerve supply. "Lat", the horizontal semicircular canal; "Post", and "Sup", the two vertical canals at right angles. The utricle (Utr.), succulus, and cochlea are labelled (after Hardy, 1934).

The external ear serves principally to channel the sound waves into the auditory meatus.

The middle ear receives the sound waves on a diaphragm (the tympanic membrane) and transmits them mechanically to the cochlea by means of a second diaphragm: the oval window.

The sound waves, reduced in intensity, are thus transmitted into the inner ear, which consists of a cavity filled with liquid. Here the vibrations mechanically and selectively stimulate the cells of the organ of Corti.

These cells pass the signals they receive to the cells of the ganglion of Corti, which then sends impulses to the brain by means of the acoustic nerve.

Connected with the internal ear are other liquid-filled cavities: the utricle, succulus and semicircular canals, which are disposed in the three planes of space.

The utricles contain mechanical-receptors which transmit information to the brain on the position of the head in relation to gravity. The semicircular canals, contain sensory cells which determine the direction and the velocity of the movement of the head.

It is obvious that an extremely mobile animal like a primate needs a very efficient sense of movement and posture, Furthermore hearing reaches a very high discriminatory level, which permits man to perceive the complex modulations of language and music. This increased power of discrimination is certainly tied to the increased complexity of the cochlea but also, and perhaps to a larger extent, to the greater evolution of the auditory cortex (consequent to the use of articulate speech) where the acoustic signals that arrive are analysed. The auditory cortex is found in the temporal lobe. In this same region, as we shall see, are also localized memory, the centres of language and the centre for perception of images.

We shall now see how the evolution, or at any rate the differentiation, of these organs in the primate series has led to a specialization and a complication of the various cerebral areas. We shall begin with the olfactory bulb.

The olfactory bulb and the pyriform lobes are concerned with the analysis of olfactory stimuli from the nose. The pyriform lobes are composed of part of the "palaeopallium", which is the most archaic part of the cerebral cortex. The palaeopallium is separated from the "neopallium" (which is posterior to it) by the rhinal sulcus. Among other functions the neopallium generally assumes the analysis of sensory impulses, except those coming from the olfactory organ, for which the palaeo- and archipallium continue to be responsible. As has been pointed out, in the evolution of primates the sense of smell decreases in importance, and with it the relative dimensions of the archipallium and palaeopallium. With the increase in importance of the other senses the neopallium expands, the rhinal sulcus extends under the sides of the encephalon, and the neopallium, including the pyriform lobe, comes to be disposed below the encephalon instead of in front of it (see Fig. 20.5).

We have also noted that the analysis of visual images is relegated to the occipital (posterior) lobe of the brain; the cortical strata of this region are therefore called the "visual cortex". It is remarkably expanded in a "lunate sulcus" or, as it is often and more appropriately called, the "simian sulcus". The development of the visual cortex is, in fact, one of the most typical characteristics of the primate brain.

The temporal lobe has also undergone enormous development since the first stages of the evolution of the primates. However, only a small part of its surface makes up the true auditory cortex (Fig. 20.8); all the rest appears to be devoted to the elaboration of the stimuli originating in vision and in the audio-visual memory. These functions are obviously very important for a group of animals whose life is firmly bound to visual memory, rather than to the memory of other sensations. Moreover, in man the temporal region acquires a further significance because the memory of many experiences are stored there.

Throughout the evolution of the primates the frontal and parietal lobes have also been greatly expanded. This part of the cortex has been clearly separated from the others during the ultimate steps in the evolution of man. It is actually the presence of the enlarged frontal lobe which has given to our species, in its final evolutionary stages, a neurocranium which extends above the orbits, and which has led to the formation of a vertical forehead. The parietal lobes, which lie behind the frontal lobe, have also been greatly enlarged. But the surface of the cortex in our species underwent a later

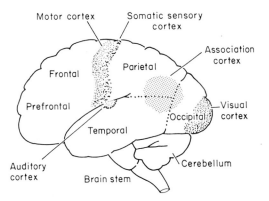

FIG. 20.8 View of the left side of the human brain, showing the main sub-divisions of the cerebral cortex, together with the somatic sensory and motor cortexes. The dotted area in the inferior posterior parietal region shows very approximately the "association cortex of association cortexes" discussed in the text (from Campbell, 1966).

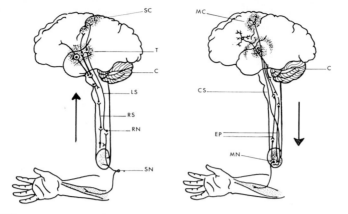

FIG. 20.9 Diagram of sensory (ascending) nerve fibres. LS labels a fibre of the lemniscal system, which forms a relatively direct link between the sensory neuron (SN) (from proprioceptors in the muscle shown here, but also from the skin) and the somatic sensory cortex (SC) via relay neurons (RN) in the brainstem near the cerebellum (C) and in the thalamus (T). Rs labels a reticular fibre that constitutes a relatively indirect link between the sensory neuron and the cortex. Diagram of motor (descending) nerve fibres. CS labels a corticospinal fibre that in higher primates forms a direct link between the motor cortex (MC) and the motor neuron (MN) of the spinal cord. EP labels two of the extrapyramidal fibres that connect the cerebral cortex through a complex sequence of relay neurons in the brainstem to the motor neurons of the spinal cord (redrawn after Noback and Moskowitz, 1963).

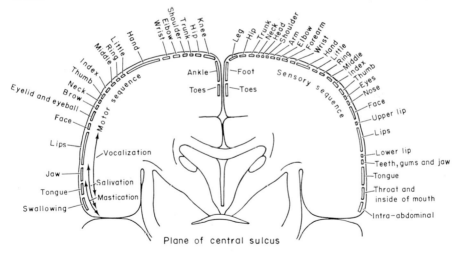

FIG. 20.10 Schematic diagram combining a cross-section of the motor cortex (left) and the somatic sensory cortex (right) of man, in the plane of the central sulcus. The diagram shows the motor and sensory sequence of representation determined by electrical stimulation of conscious patients. Note that the sensory area devoted to the fingers and hand is greater than that devoted to the much larger skin area lying between forearm and shoulder (from Rasmussen and Penfield, 1947).

development, and therefore is folded in a complex system of sulci. In man 64% of the surface of the cortex is found folded in sulci (or fissures); in the more primitive monkeys only 7%.

The functions of the frontal and parietal lobes are only partly known. Part of the cortex situated between these two lobes is clearly identifiable as "motor cortex" and as "somato-sensitive cortex" (Fig. 20.8). In these areas the transmitting and receiving neurons of the body's motor organs and somatostatic organs (sensory receptors from the skin) are located (Fig. 20.9). This area has been mapped by means of electrostimulation.

The frontal lobes, with a prefrontal area, are located anterior to the motor cortex. Since no clear reaction has been obtained from it by means of electrical stimulation, this is called the "silent "area. The sole evidence of its function derives from the surprising and complex changes in the personalities of individuals who have lost this part of the brain through some accident. These data suggest that exercise of the will can be assigned to this region. The prefrontal area seems to be responsible for initiation and concentration of attention in human behaviour.

The capacity of man to concentrate and to direct his attention for a long time towards a particular object must have been an important achievement in his evolution. Kortlandt (1965) affirms that, as a rule, the non-human primates cannot succeed in focusing their attention on an object for more than a quarter of an hour. This limitation contrasts with the prolonged attention span of the carnivores, which follow the movements of their prey for hours, or even days, Obviously, this difference has been established because it presents a selective advantage. A herbivore may die of hunger, if he searches for special grasses, while a carnivore easily distracted from his prey, once spied, has little chance of relieving his hunger.

When the progenitor of man, beginning as a tree-dweller and vegetarian, also became a carnivore and hunter, he had to develop and concentrate his attention towards the motive behind his actions. Thus he had to develop and evolve new cerebral structures. Following this line of reasoning, we are able to see how variations in the ecology and behaviour of the ancestors of our species required the evolution of new forms of mental processes, which then came to provide the essential foundation of our culture.

Our knowledge of the function of the parietal lobes is not very great, but we must remember that their expansion has been of the utmost importance in human evolution. In terms of function the parietal areas of the cortex are usually denoted as the "cortex of association", a descriptive term which signifies that this area receives intra-cortical connections from three sensory-receptor areas: the auditory (in the temporal lobe), the somataesthetic and the visual (in the occipital lobe), related respectively to hearing, touch and sight. In the monkeys, and particularly in the anthropoid apes, these areas are directly linked to the surrounding cortex, so that the near-by areas are called the areas of auditory association, somataesthetic association and visual association. In man, on the contrary, there is an expansion of the parietal lobe between these three areas, which leads to the formation of what has been described as the "associative cortex of the associative cortices" (Geschwend, 1964). This area of secondary association is very small in the anthropoid apes, while it is enormous in man. All its connections are with the surrounding cortex rather than with the central parts of the brain.

The "associative cortex of the associative cortices" is the inferior-posterior parietal region (the shaded area in Fig. 20.8) and is of great interest because at least part of it seems to coincide with the major area concerned with articulate language in Man.

The result of the expansion of the frontal and parietal lobes in the final stages of human evolution is such that they have covered not only the archipallium (hippocampal circonvolution) but also most of the visual cortex.

In fact the simian sulcus, which delimits the visual cortex in the monkeys, particularly in the anthropoids, is almost obliterated in man. The visual cortex has come to lie, for the greater part, under the encephalon rather than above it, and is therefore difficult to identify.

The primary motor areas, the somataesthetic, visual and auditory ones, which occupy the greater part of the surface of the cerebral hemispheres of a monkey, in man are crowded into the sulci.

This increase in the cortical areas also requires an increase in the connections between the various areas. This is manifested in the enormous development of the corpus callosum, which contains the collection of fibres that link the two cerebral hemispheres (Fig. 20.12).

Other than the cerebral hemispheres, two other parts of the brain are important: the thalamus and the cerebellum. The thalamus, following the remarkable evolution of the neocortex, is found, as it were, cemented in the mass of the telencephalon. It is linked to the evolution of the neocortex and the appearance of a neocortex and the appearance of a neothalamic area, and functions as a relay station for the sensory pathways (lemniscal) that continue on into the various neocortical areas.

The cerebellum, which like the cerebral cortex has been conspicuously developed during evolution, coordinates the activity of the skeletal muscles. It controls the force and the time of every muscle contraction involved in a movement. Its activity is involuntary, in that it is not subject to the control of the cortex of the cerebral hemispheres, but is of enormous importance in the mammals in whom rapid and accurately controlled movements are necessary. A high degree of neuromuscular coordination is more essential to the primates than to any other mammal, because of their arboreal life.

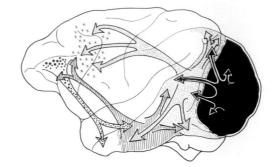

FIG. 20.11 Diagram of some of the intracortical connections in the brain of the monkey *Macaca* recently revealed by Kuypers. Note the dense connections between the visual cortex (black), its neighbouring association cortex (shaded), and the temporal lobe (after Kuypers, 1965).

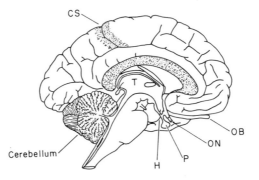

FIG. 20.12 Median sagittal section of the human brain showing (upper part) the internal (medial) surfaces of the left cerebral hemisphere, below which lie the transverse fibres of the corpus callosum that connect the two hemispheres (shaded). The central sulcus (CS) separates the somatic sensory and motor cortexes (shaded) which continue on the median side of the hemisphere. The left olfactory bulb (OB) and optic nerve (ON) are shown at the base of the brain; together with the thalamus (T), hypothalamus (H), and pituitary (P). Like cerebellum and the brainstem, these organs lie medially and are cut in this view (from Campbell, 1966).

Man has inherited a highly evolved cerebellum from the non-human primates. Without any doubt this has been of the greatest consequence, not only in the acquisition of bipedal locomotion and upright posture, but also in the delicate motor control necessary for the making of tools.

From this brief synthesis we can envisage how the different parts of the brain developed in a continuous succession of steps, which have led from *Adapis* to Man. Presumably each expansion was in relation to the new functional needs of the evolutionary stage in which it was acquired, and each played its rôle in the creation of *Homo sapiens*.

Among these, the development of the temporal and parietal lobes seems to have created heightened possibilities for man to integrate his experiences, since the frontal lobe gave him new control over his behaviour. Nevertheless it is not the brain that has brought about the evolution in human behaviour. The brain of man has been differentiated, together with his behaviour, in response to environmental change.

General References

Bianchi, L. (1968). Evolution of the brain in the Old World Primates. *In* "Taxonomy and Phylogeny of Old World Primates" (B. Chiarelli, ed.). Rosenberg and Sellier, Turin.

Brothwell, D. (1963). Where and when did man become wise? *Discovery* **2**/3, 10–4.

Campbell, B. G. (1966). "Human Evolution. An Introduction to Man's Adaptations". Aldine Publishing Company, Chicago.

Heines, D. E. and Swindler, D. R. (1972). Comparative neuroanatomical evidence and taxonomy of the Tree Shrews (*Tupaia*) *J. Human Evol.* **1**, 407–420.

Kortlandt, A. (1965). Comment on the essential morphological basis for human culture. *Curr. Anthropol.* **6**, 320–325.

Kuypers, II. G. (1964). The descending pathways to the spinal cord, their anatomy and function. *In* "Progress in Brain Research" (C. Eccles and P. Shade, eds), Vol. 2, pp. 178–202.

Penfield, W. and Roberts, L. (1959). "Speech and Brain Mechanisms", Princeton University Press, Princeton, N.J.

Tobias, P. (1970). "Evolution of Hominoid Brain", Columbia University Press, New York.

Von Bonin, G. (1963). "The Evolution of the Human Brain". The University of Chicago Press, Chicago.

Epilogue

The living non-human Primates are of great importance to anyone wishing to relate the study of the biology and evolution of man to its proper context. Whether we are concerned with molecular biology, immunochemistry, cytogenetics, physiology, anatomy or pathology, it is impossible to remain unaware of the extremely interesting discoveries that have been made recently on non-human Primates. The 180 species of the existing Primates provide a wonderful experimental material for comparative studies which certainly will provide important conclusions of the greatest interest not only for general biology and evolution, but also for the applied biology of our own species: especially medicine.

Unfortunately, at the very moment when we are becoming aware of the uniqueness of the non-human Primates, we also realize how precarious their future is and how competition with industrial man is threatening their survival. Some of the species listed fifty years ago do not exist any more and at least half of the 180 actually existing ones including the great Apes are in extreme danger of extinction. It is certainly not of little interest to humanity to allow these lower relatives to disappear. Can we remain unmoved at such annihilation?

Tomorrow, more than today, scientists will need the non-human Primates to solve important problems of human health and knowledge. We therefore must do our best now to propose a conservation policy to the international and national authorities concerned to ensure the survival of substantial populations of the various forms of the present Primates.

An appeal for conservation of non-human Primates was launched at a NATO Advanced Study Institute on Comparative Biology of Primates in June, 1972. Here is the text for those who are concerned.

Appeal for Conservation of Non-human Primates

The scientists of many disciplines and different countries participating in the NATO Advanced Study Institute on Comparative Biology of Primates, held in Montaldo (Torino), June 7-19, 1972:

> being aware of the unique value of non-human primates as man's closest relatives and as models serving the biological and medical sciences in the advancement of human health, welfare and knowledge;

being aware of their responsibility to preserve the existence of the whole spectrum of contemporary primate species;

being aware that expanding human populations and the growing usage of non-human primates have threatened some species with extinction while others have become drastically reduced.

URGENTLY SUBMIT to International and National Organizations the following appeal that:

1. Scientists be selective in the usage of non-human primates and employ other animal models when they are appropriate, and require that endangered and rare species be limited to investigations in which other species of non-human primates are unsuitable.

2. Scientists contribute to the conservation needs of non-human primates by:

introducing and insisting upon, humanitarian and efficient procedures for their capture, translocation and maintenance prior to and during their use;

educating the public about the real health hazards presented by non-human primates which carry many transferable diseases (especially viral infections), and recommending that the use of monkeys and apes as pets be prohibited.

insisting upon methods of acquisition which ensure the enforcement of national laws covering capture and transportation and the international support of these laws;

promoting the development of knowledge on the distribution and status of non-human primate populations in the areas where they occur and of management and husbandry methods which ensure their survival in natural ecosystems and as economic and scientific resources;

promoting the development of permanent breeding programmes according to a long term requirement for different species;

encouraging the urgent cooperation, financial support and investments from research institutions, pharmaceutical and other industries to accomplish these aims for ensuring the continued existence and development of non-human primate populations in their natural ecosystems, or in especially designed environments, so that their availability for human use is ensured for posterity.

The appeal has been approved and accepted with few modifications by the International Primatological Society on its IV meeting in Portland on August 1972.

Subject Index